MATH CATS

MATH CATS

Scratching the Surface *of* Mathematical Concepts

DANIEL M. LOOK, PhD

RUNNING PRESS
PHILADELPHIA

Running Press
Hachette Book Group
1290 Avenue of the Americas, New York, NY 10104
www.runningpress.com
@Running_Press

First Edition: October 2025

Published by Running Press, an imprint of Hachette Book Group, Inc. The Running Press name and logo are trademarks of Hachette Book Group, Inc.

Print book cover and interior design by Tanvi Baghele

Library of Congress Cataloging-in-Publication Data has been applied for.

ISBNs: 979-8-89414-000-1 (hardcover), 979-8-89414-002-5 (ebook)

Printed in China

APS

10 9 8 7 6 5 4 3 2 1

To my family:

My incredible wife, Laura, who is always on board with my ridiculous ideas.

My big brothers, Alfred and David, who I always looked up to.

My mom and dad, who inspired me and taught me the importance of laughter and joy in life.

CONTENTS

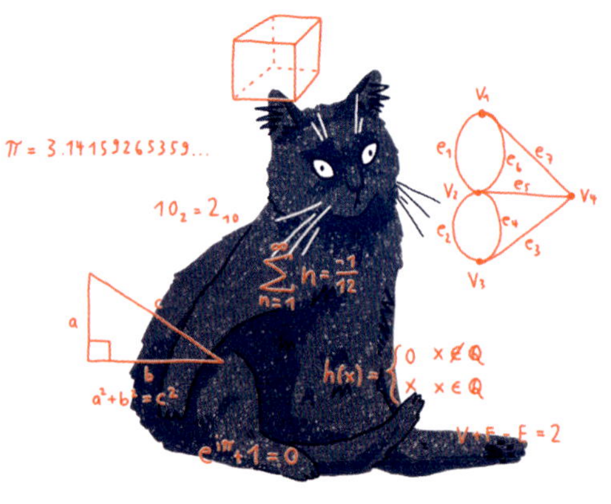

$\pi = 3.14159265359...$
$10_2 = 2_{10}$
$\sum_{n=1}^{\infty} n = \frac{-1}{12}$
a
b
c
$a^2+b^2=c^2$
$h(x) = \begin{cases} 0 & x \notin Q \\ x & x \in Q \end{cases}$
$e^{i\pi}+1=0$
V_1
V_2
V_3
V_4
e_1
e_2
e_3
e_4
e_5
e_6
e_7
$\epsilon > 0$

INTRODUCTION

When my roommate Meow Meow first approached me about doing this collaboration, I admit I was hesitant. Don't get me wrong, math is fascinating and weird (sometimes *really* weird), and cats are just darn adorable. However, I didn't have a vision for how to combine these topics. I mean, don't cats and math go together about as well as cats and dogs? But there was something about the look in his eyes and the twitch in those whiskers that made me reconsider.

And then it kind of made sense. I know I've sometimes wished that I had nine lifetimes to master a subject. And who has nine lives? That's right. Cats.

The following are snippets from conversations among myself and Meow Meow and the other cats we live with—SpaghettiOs, Conan T. Cat, and Mrs. Waffles. The intent was to have fun with some interesting mathematics, choosing examples that resonated with my cat companions.

We avoided topics requiring a lot of prerequisite knowledge (cats are not fans of hierarchical learning) and, as much as possible, the math notation that many people find intimidating and cats refuse to acknowledge. We kept the entries short, but more details on the concepts can be found online. Or you can ask a cat.

PURRTHAGOREAN THEOREM

Triangles (and Cults)

Perhaps you've heard of Pythagoras, the Greek philosopher, polymath, and possible onetime cult leader (yeah, we'll touch on that later) and how he did some things with triangles and had a theorem. But have you heard of Purrthagoras and the Purrthagorean Theorem? Since triangles can be boring and everything is better with cats, let's talk about the Purrthagorean Theorem!

When Beans holds his front legs at right angles, the distance between his front paws is called his *hypawtenuse*.

The Purrthagorean Theorem gives a relationship between the length of a cat's two front legs and their hypawtenuse. Specifically, the sum of the squares of the two leg-lengths equals the square of the hypawtenuse. Denoting the length of the left and right front legs as a and b, respectively, and the length of the hypawtenuse as c we get the famous equation $a^2 + b^2 = c^2$.

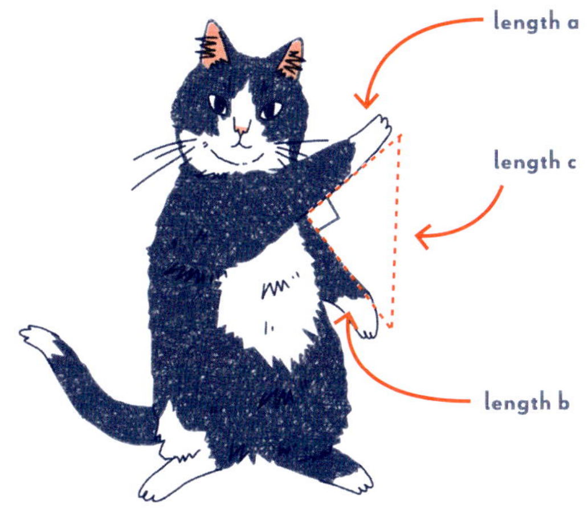

In our case, Beans's front legs are each 8 inches long. Then $a = b = 8$. So, what is c, the hypawtenuse? Well, $8^2 + 8^2 = 128$, meaning $c^2 = 128$, implying that c is the square root of 128, or roughly 11.31 inches. If Beans holds his front legs at right angles, there will be approximately 11.31 inches' distance from his left front paw to his right.

The theorem is named after Pythagoras of Samos (ca. 570–495 BCE), a Greek philosopher and mathematician. Take for a fact that Pythagoras was sharp. However, most of his life is blurred by legend. He is credited with many discoveries, and the famous Pythagorean Theorem bears his name, yet it is unclear how many of the discoveries attributed to him are actually due to him. In fact, some historians dispute whether he personally even advanced the field of mathematics.

For example, the Egyptians knew the Pythagorean Theorem at least one thousand years *before* Pythagoras was even born. (Pythagoras traveled through Egypt and may have learned of the relationship there.) Making it worse, some results attributed to him may have been discovered by members of his sect, the Pythagoreans. And then there is the debate over whether the Pythagorean sect should really be called the Pythagorean cult. (No, really, some do consider them a math cult; look it up!)

We won't weigh in on the cult status. Instead, we'll avoid controversy and just say that out of the two sects, I prefer the Purrthagoreans, who pretty much believe only in math and cuddles.

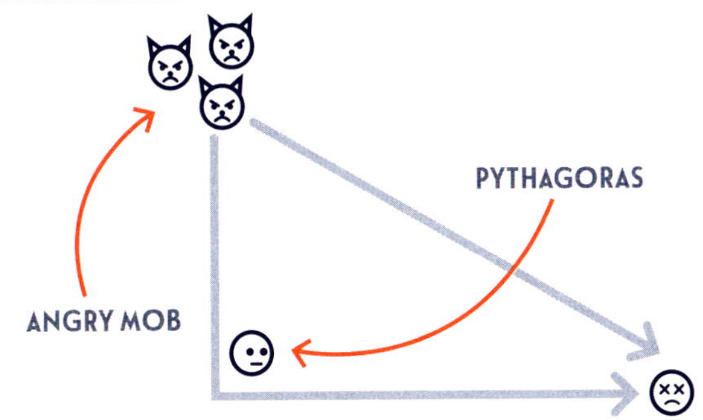

The Pythagoreans had some interesting quirks. For one, it reportedly took five years of initiation to join. They also couldn't eat meat. But, more curiously, the Pythagoreans were prohibited from eating (or possibly even touching) beans. One explanation for this rule was that Pythagoras believed beans contained the souls of the dead due to their fleshlike consistency and shape. But the beans lore doesn't stop there!

One account of Pythagoras's death entails being chased by a mob of anti-Pythagorean arsonists. Pythagoras refused to run through a field of fava beans and as result was caught and killed by the mob. Imagine Pythagoras running toward a line of beans, then banking left, while the mob ran the diagonal to catch him. Pythagoras was most likely the only mathematician whose death can be explained using a theorem named after him.

WAIT, THERE ARE CATS THAT AREN'T ACUTE?

Angles and Degrees

The following cats are all cute, but only one is acute...and another is obtuse (no offense), while the third is right (and never lets you forget it). So, which is which?

MARMALADE

SPAGHETTIOS

PANCAKE

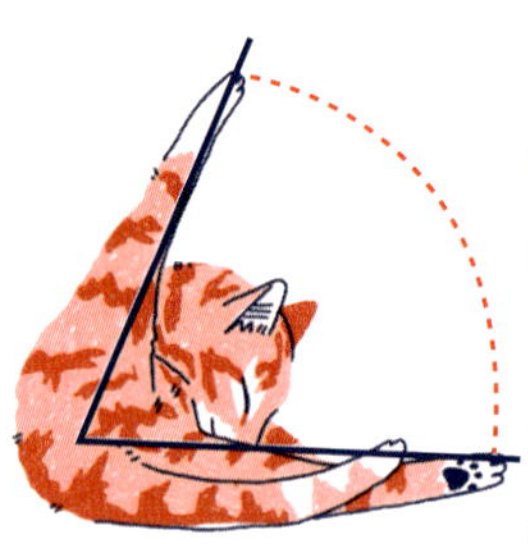

You may have recognized that the terms *acute*, *right*, and *obtuse* all refer to types of angles. An acute angle is an angle between 0 and 90 degrees. Marmalade's legs make an acute angle during his morning hygienic routine.

An obtuse angle is one whose measure is between 90 and 180 degrees. This is demonstrated by SpaghettiOs during her stretch.

SPAGHETTIOS IS A LITTLE TORTIE.

Finally, a right angle is one that is precisely 90 degrees. Mr. Daryl Pancake's show-cat posture contains a right angle.

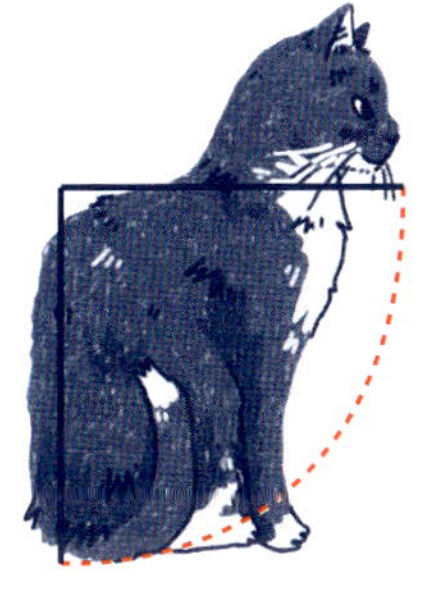

But what if I told you that, of these three angle types, the *right* angle was the first named...and when it was described there was no mention of degrees at all? In fact, there wasn't even a unit of measure for angles at that time. So, how were right angles defined? Why do we now use degrees for angles? Why are there 360 degrees in a circle? And what *is* a degree, other than temperatures and things graduate students often bring up?

In the *Elements* (ca. 300 BCE), Greek mathematician Euclid (ca. 330–270 BCE) defined a *right angle* as an angle created from the intersection of two perpendicular lines.

> *When a straight line standing on a straight line makes the adjacent angles equal to one another, each of the equal angles is right, and the straight line standing on the other is called a perpendicular to that on which it stands.*
>
> —*Elements*, Book I, Definition 10

This term *right angle* likely derives from the Latin descriptor *rectus*, which means "straight," but also refers to perpendicular. The terms *acute* and *obtuse* were introduced much later, with acute referring to "sharp" and obtuse to "blunt."

Using 360 divisions, or degrees, comes from ancient Babylonian and Sumerian astronomy. The exact reason for 360 is unknown, but there are several competing theories. One theory is based on calendars and celestial motion. The Babylonians had more than one calendar, including an "ideal calendar" used primarily for administrative work and astronomy. This calendar consisted of 12 divisions with 30 subdivisions each. This gives a total of 360 subdivisions, which roughly correlate with days, giving a 360-day year. (Meanwhile, a cat's ideal calendar would likely be 360 subdivisions of scritches, napping, and eating.)

How does this help us *define* a degree? Consider three objects: the earth, the sun, and a fixed constellation, Purrseus (known as Perseus in English). Imagine that there is a time during the year when these three objects are in a line. If you were able to see Purrseus during the day *and* if you could safely stare straight at the sun, then it would look like the sun was directly in front of Purrseus. *Do not stare straight into the sun.*

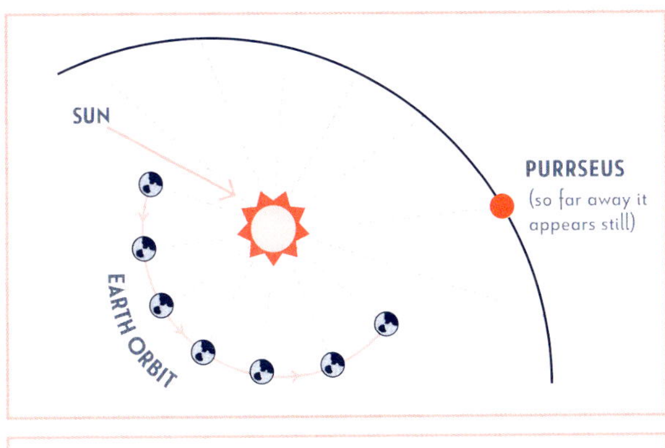

The earth revolves around the sun, the sun itself is moving, and so is Purrseus. If you return at the same time the next day, and again stare at Purrseus, the sun will appear to have moved a little bit. This distance was defined as a *degree*.

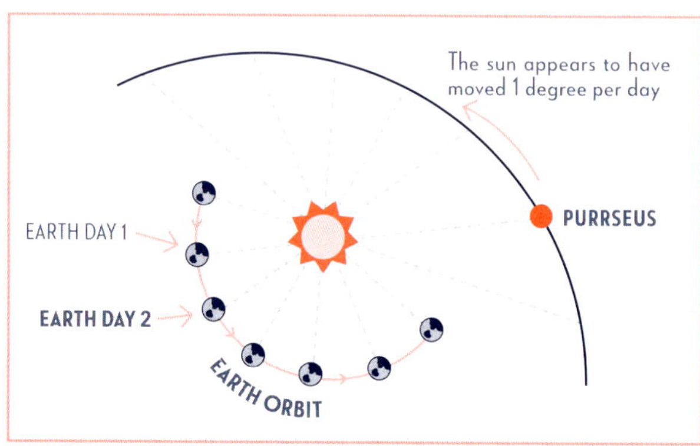

DAY 2

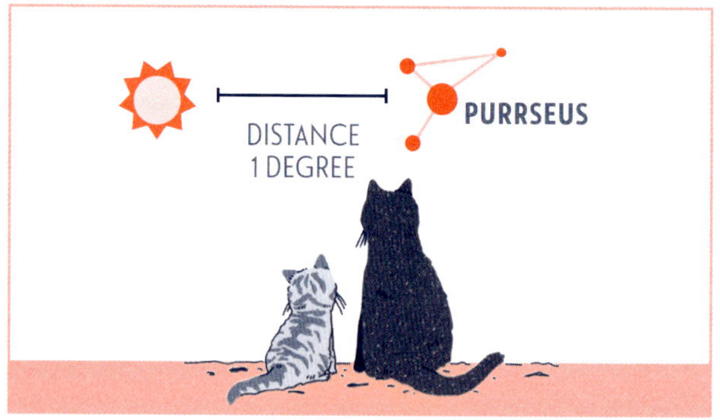

After 359 days, the sun will once again be 1 degree from Purrseus, but now moving closer.

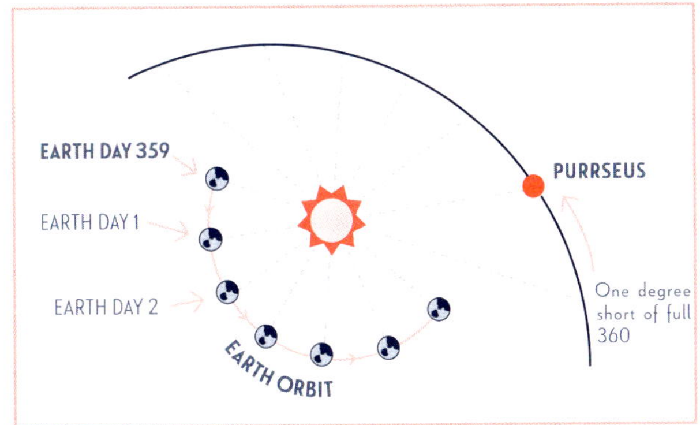

DAY 359

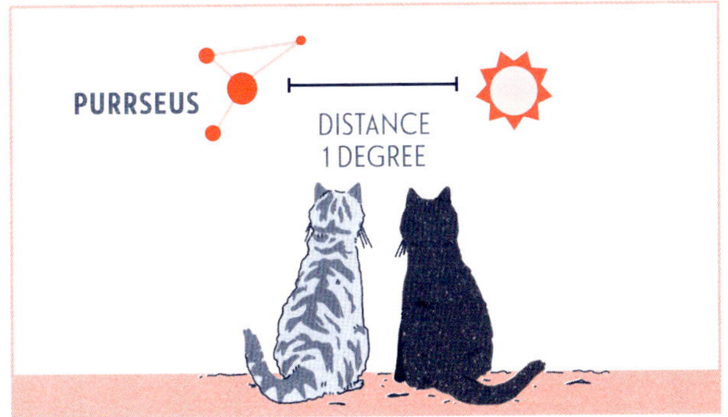

After a full year, the sun will return to its position in front of Purrseus.

The path of the sun along the ecliptic is kind of circular, and the Babylonians divided it into 360 subdivisions, or degrees. Now fast-forward...a lot. Hipparchus of Rhodes (ca. 190–120 BCE) applied geometry to Babylonian astronomy, and he needed a way to measure the magnitude of angles. As the Babylonians had, essentially, provided a division of circles into 360 units via the path of the sun, it was natural to use the Babylonian degree to assign values to angles.

Tater Tot, in partial fulfillment of the requirements for her MFA (master of the feline arts), wrote a stirring thesis, "Of Revelations & Revolutions," describing Hipparchus's work as relayed by his cat.

The "ideal calendar" was not used in day-to-day life. The Babylonians were very aware that a year did not have exactly 360 days. The 360-day calendar was used only for fiscal or astronomical events. There was also a lunisolar calendar that consisted of 12 months, with a 13th month inserted as needed to align the calendar with the seasons, like the extra day in a leap year.

DIVIDE LIKE AN EGYPTIAN

Egyptian Fractions

Even in ancient times, cats were cats. Here, Pickles (translated from ancient Egyptian) is pushing a statue over a ledge, breaking it into six pieces. However, in this case there is a deeper message about . . . fractions?

The statue Pickles broke represents the Eye of Horus. It has been suggested that ancient Egyptians associated the pieces of the eye with certain fractions. This is most likely not true, but we can use the idea to learn a bit about how ancient Egyptians wrote fractions.

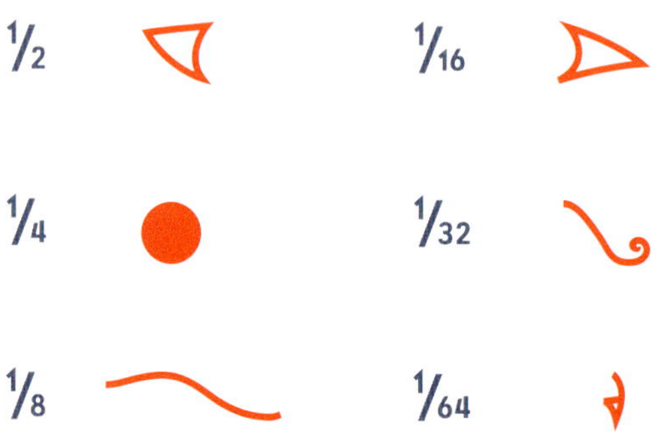

Ancient Egyptians had a deep understanding of positive rational numbers, or what most of us would call positive fractions, which we'll just call *fractions* here. When you hear *fraction*, you probably think about numbers like $\frac{1}{3}$, $\frac{5}{9}$, and $\frac{100}{8}$. Basically, $\frac{a}{b}$ where *a* and *b* are a couple of whole numbers and *b* can't be zero. (Whole numbers are the numbers 0, 1, 2, 3, . . .)

The fractions ancient Egyptians worked with came in a specific form. They were fractions with ones in the numerator (the top of the fraction). They could also be sums of such fractions. So, $1/2$ and $1/50$ are in what we'll call Egyptian notation, as is $1/13 + 1/22 + 1/213$, but $2/3$ is not.

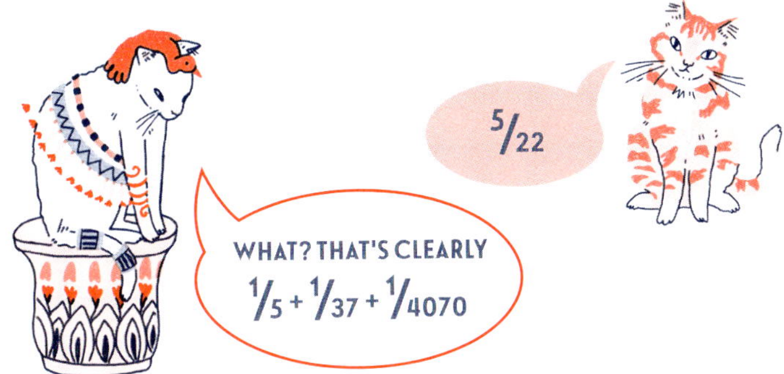

This does not mean that the Egyptians didn't know about $2/3$. They did. However, they would have called it $1/2 + 1/6$. This sum equals $2/3$. The distinction is kind of like using "we'll" instead of "we will." When a fraction is expressed as a sum of fractions with ones on top, we'll say it is an *Egyptian fraction*. Having two ways to refer to the same number is similar to binary on page 56.

Our fast-and-loose fractions, where we don't require ones on top, are called *vulgar fractions*. (Gotta admit, I love that name.) And it has been proven that any fraction can be written in either Egyptian or vulgar form, which is kind of neat...You can always decompose a (positive) fraction into a sum of distinct fractions with ones in the numerator. For example, $5/22 = 1/5 + 1/37 + 1/4070$.

Egyptian fractions were used for a long time; Fibonacci's 1202 book *Liber Abaci* (*The Book of Calculations*) has sections devoted to Egyptian fractions. (For more about Fibonacci, see "The Fibonacci Sequence" on page 19.)

However, writing fractions this way is clunky and was surpassed by our current notation, based on a sixth-century Indian version of fractions with some upgrades from the twelfth-century Moroccan mathematician Al-Hassãr. (He's the one who added the little bar in the middle. Before that, a fraction was just one number floating above another.)

Although not easy to work with, there are some advantages to Egyptian fractional notation. What if you have six cats but only five cans of cat food? Clearly, each cat gets ⅚ of a can, right?

True, but that is a pain. We need to pull out $\frac{5}{6}$ of the first can, leaving $\frac{1}{6}$...then we need to add another $\frac{4}{6}$ to get the second helping of $\frac{5}{6}$. This could be hard.

But if we rewrite $\frac{5}{6}$ using the Egyptian convention, we have $\frac{5}{6} = \frac{1}{2} + \frac{1}{3}$. This is much easier! Each cat gets $\frac{1}{2}$ of a can along with $\frac{1}{3}$ of a can. Since we need six servings of $\frac{1}{2}$ of a can, we divide three cans in half. We then have two cans left over, each of which we divide into thirds, giving six servings of $\frac{1}{3}$ of a can.

Giving all kitties $\frac{1}{2}$ of a can along with $\frac{1}{3}$ of a can divides the food equally. Bam! Happy kitties!

The fractions supposedly linked to the pieces of the Eye of Horus are all Egyptian fractions that come from dividing by two repeatedly. Start with $\frac{1}{2}$. Dividing by two gives $\frac{1}{4}$, then $\frac{1}{8}$, and so on. These are called *heqat fractions.* The heqat is a unit of measurement for wheat, and the heqat fractions come from taking successive halves of the heqat. Kind of like how cats usually buy their milk in gallons or half gallons.

In a version of the myth of Horus's eye, his eye is torn out and the pieces are scattered. Yikes. But he's got grit, so he collects the pieces and puts them back together. Unfortunately, something was missing...

If you add the fractions associated with his eye parts, you get $\frac{1}{2} + \frac{1}{4} + \frac{1}{8} + \frac{1}{16} + \frac{1}{32} + \frac{1}{64} = \frac{63}{64}$. We are shy $\frac{1}{64}$ of an eye. Horus may have left a piece out, like the missing screw in the box for Mr. Skrittles's Swedish cabinet.

FURBONACCI CATS

The Fibonacci Sequence

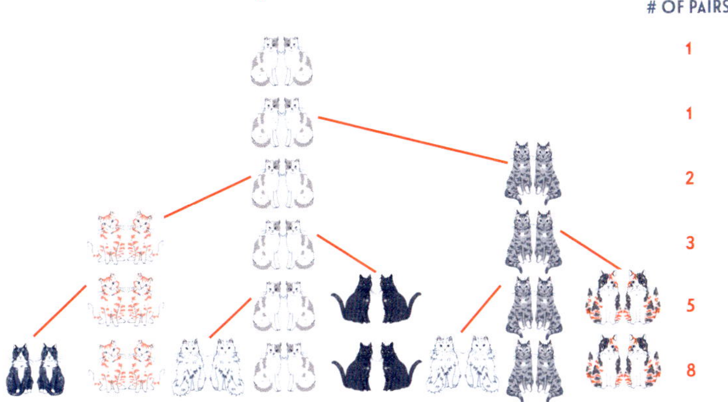

Let's say you have two cats, and—hypothetically—they could have a pair of kittens one month after being born. How many pairs of cats would you have in a year? Or ten years?

No need to lose sleep over this question. We've had the formula since 1202, when it was featured in *Liber Abaci* by mathematician Fibonacci (1170–ca. 1250).

Interestingly, Fibonacci was not known as Fibonacci during his lifetime; this name was given posthumously by historian Guillaume Libri in 1838. He was known by a few names, but he wrote *Liber Abaci* as "Leonardo, son of Bonacci, of Pisa" (Bonacci refers to his family name). The name Fibonacci was derived from "son of Bonacci" in Latin, *filius Bonacci*. For simplicity and sanity, we're respectfully just going to call him Fibonacci.

In *Liber Abaci*, Fibonacci discussed, among other things, rabbit populations with an unrealistic life cycle described by the following assumptions:

1. **It takes one month for a pair of baby rabbits to become a pair of adult rabbits.**
2. **Each month, every pair of adult rabbits produces a single pair of baby rabbits.**
3. **The rabbits never die.**

Fibonacci's concept was inherently flawed, as any outdoor cat worth a damn can kill a rabbit, so we're just going to use cats to figure this out.

Through *Liber Abaci*, Fibonacci also introduced Europeans to the Hindu-Arabic numeral system (our current base-10 number system; see "Binary" on page 56 for more details) along with the symbols for 0 through 9. Until this time, Europeans were still using Roman numerals. In fact, one of the purposes of *Liber Abaci* was to convince people, especially merchants, to abandon Roman numerals in favor of the Arabic enumeration.

Move over, Fibonacci; Furbonacci is taking it from here.

In mathematics, an infinite string of numbers is called a *sequence*, and listing the number of pairs of cats that we are trying to feed each month gives one of the most famous sequences of all—you guessed it, the Fibonacci—or, as we'll call it, the Furbonacci—sequence: 1, 1, 2, 3, 5, 8,...

In the first month, you have a pair of kittens.

In the second month, you still have one pair, but now they're adult cats.

In the third month, you have two pairs—the original and their newborn kittens.

In the fourth month, the original pair has a second pair of newborns, and their first kittens have grown up, so three pairs total!

You may have noticed an interesting pattern to the sequence (in this case, a recursive pattern). Each new number of pairs is the sum of the previous two. So, 1 + 1 = 2, 2 + 1 = 3, 3 + 2 = 5, 5 + 3 = 8, and so forth.

So how many pairs of cats will we have in a year?

That is an astounding 144 pairs of cats! I suggest getting your little ones spayed and neutered, although my cats vehemently disagree.

MONTH	KITTEN PAIRS	ADULT CAT PAIRS	TOTAL PAIRS
1	1	0	1
2	0	1	1
3	1	1	2
4	1	2	3
5	2	3	5
6	3	5	8
7	5	8	13
8	8	13	21
9	13	21	34
10	21	34	55
11	34	55	89
12	55	89	144

As for how many pairs of cats you'll have in ten years, well, don't expect me to do the math for you. Follow the sequence; Furbonacci will never steer you wrong.

Although this sequence is named after Fibonacci, it was known well before Fibonacci's time. As early as 200 BCE, Indian poet and mathematician Acharya Pingala found this sequence while counting the total number of possible patterns of Sanskrit poetry formed from syllables of two different lengths. Naturally, cats, being fans of poetry (and all literary arts), are well aware of this.

CURL UP WITH MATH

The Fibonacci Spiral

A catnap becomes cat math when Pumpkiccino curls up in this common position, taking the shape of a spiral—in this case, the *Fibonacci* (or *Furbonacci*) *spiral*. As you may have guessed by the name, this spiral will involve the Fibonacci sequence. (See "The Fibonacci Sequence" on page 19.)

To construct this spiral, start with two squares, each with side length one, sharing a common side:

FIG A

In the upper square, draw a quarter circle of radius one:

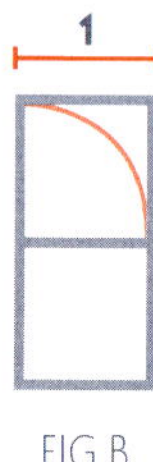

FIG B

Adjoin a square of side length two:

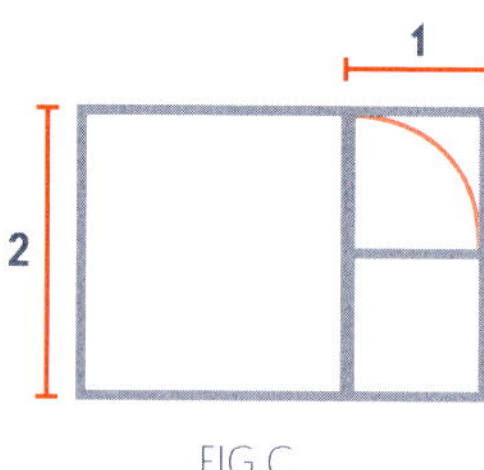

FIG C

Now put a quarter circle of radius two in the square of side length two:

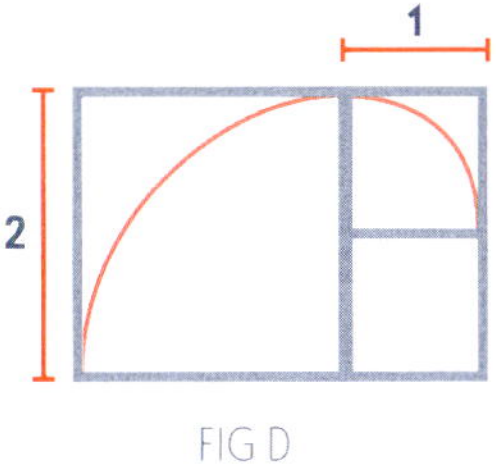

FIG D

Continuing this pattern, we would add a square of side length three to the bottom, then a square of side length five to the right side, and so on. (Here is where we see the Fibonacci sequence: 1, 1, 2, 3, 5, 8, . . .) This is the image we get when we stop at the square of side length five:

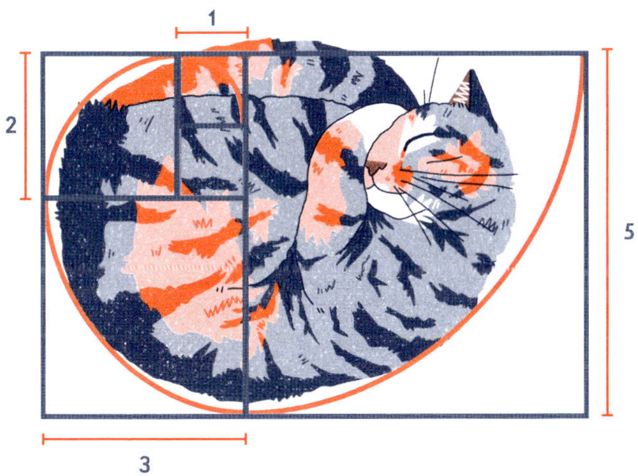

If we have a lot of time on our hands and continue this forever, the curve we obtain is the Fibonacci spiral.

Beyond being Pumpkiccino's favorite sleeping position, what is so special about this spiral? First, it approximates a mathematical object known as the *golden spiral*. In the Fibonacci spiral, the spiral grows in discrete steps, going from radius one to two and then two to three as it passes through the corners of the squares. The golden spiral grows *continuously* by a factor of the golden ratio, which is represented by the Greek letter φ. (See "The Golden Ratio" on page 28 for more details.) If we are some distance from the center of the spiral and we follow the spiral for a quarter turn (that is, 90 degrees), the distance from the center of the spiral will have increased by a factor of φ.

The golden spiral is an example of a *logarithmic spiral*, and, in addition to Pumpkiccino's sleeping pattern, these spirals are known to appear in nature. The shell of a nautilus, for example, is approximately a logarithmic spiral. In fact, a lot of natural objects have similarities to logarithmic spirals...examples include the flight path of certain birds of prey when approaching prey, nerves in your cornea, and even spiral galaxies.

Although not Fibonacci spirals, sunflowers contain another connection between spirals and the Fibonacci numbers. The next time you see a sunflower, look at the seed pattern. It contains clockwise and counterclockwise spirals. If you count the number of spirals in each direction, you will always get two consecutive Fibonacci numbers. The same happens with pine cones. Pineapples have three different spiral patterns that can be counted, and you will always get three consecutive Fibonacci numbers.

So, when cats curl into little balls, they are doing more than being adorable; they are showing us a marvelous connection between mathematics and nature.

In the Fibonacci spiral, the quarter circles grew from radius one to two (so a rate of $^2/_1$), then from two to three (rate of $^3/_2$), then from three to five (rate of $^5/_3$), and so on. Essentially, the growth rate can be thought of as the sequence created by dividing consecutive Fibonacci numbers: $^2/_1$, $^3/_2$, $^5/_3$, $^8/_5$, $^{13}/_8$, . . . Turns out, this sequence converges to φ, which is why we say the Fibonacci spiral approximates the golden spiral.

THE OBJECTIVELY CUTEST CAT

The Golden Ratio

All cats are cute...but is there a formula to help us determine exactly *how* cute? It turns out that a study has been conducted to answer just this question using the mathematical concept of the *golden ratio*. So, what is the golden ratio, and what does it have to do with the adorableness of cats?

To define the golden ratio, we start with a line segment divided into two lengths: A and B.

From these two lengths, we consider two ratios, which are essentially fractions. One of these is the ratio of the longer segment to the shorter segment, or A/B. The other is the ratio of the entire length to the longer segment, giving us $(A+B)/A$. There are many values of A and B where these two ratios are equal, but whenever they are equal, the shared value is always approximately 1.618 (or to be exact, $\frac{1+\sqrt{5}}{2}$). This value is called the golden ratio, and it is denoted by the Greek letter φ.

Consider Snickers. The length from Snickers's neck to the base of her tail is 50 centimeters, while her tail is 31 centimeters long. Therefore, $\frac{\textit{longer}}{\textit{shorter}} = \frac{50}{31}$ and $\frac{\textit{total}}{\textit{longer}} = \frac{81}{50}$. Since $50/31$ is roughly 1.6129, while $81/50$ is exactly 1.62, we see that these values are not equal. So, the ratio of body length to tail length for Snickers, although close, is not the golden ratio. Poor Snickers!

The golden ratio has been studied under many different names by mathematicians for thousands of years. The ancient Greeks became aware of the importance of this value through its repeated appearance in geometry. Greek mathematician

Euclid (first mentioned in "Angles and Degrees" on page 6) referred to this value as the *extreme and mean ratio*. It was not until 1835 that this value was described as "golden" in a book by German mathematician Martin Ohm (1792–1872).

What does any of this have to do with measuring Snickers's visual beauty? The notion that the golden ratio is connected to aesthetic appeal has been around for a long time. A quick internet search yields claims that the golden ratio was intentionally used in, among other examples, the design of the pyramids of ancient Egypt, the famed Parthenon, and Leonardo da Vinci's *Vitruvian Man* and *Mona Lisa*.

Although the above examples are either disputed or have been refuted outright, that doesn't mean the golden ratio hasn't been used in art and architecture for aesthetic purposes. Known cat lover Salvador Dalí's *The Sacrament of the Last Supper*, for example, uses a canvas whose side lengths approximate the golden ratio, and French architect Le Corbusier's system of proportions, known as the modular man, was also based on the golden ratio. However, these uses came after the urban legend of the golden ratio in classical art and architecture.

Regardless, truth never stood in the way of proclamations of ultimate cuteness in cats, so if you have a golden kitty, there is no shame in sharing that fact.

Researchers at Cats.com used various measurements across cats to see which breeds came closest to the golden ratio, publishing the results in the story "Beautiful Cats: The Golden Ratio Reveals the Most Beautiful Cat Breeds." If you are curious, there was a three-way tie for first place between the Norwegian Forest, Russian Blue, and Manx breeds.

CALCULATE BEFORE YOU LEAP

Euler's Method

Cats excel at parkour, executing leaps with amazing accuracy... at least, most of the time.

Below is a pretty common sight. Butterball sees a plant and wants to get to the plant. The only rub? She needs to execute a jump.

She could use intuition to figure out how she should jump, but what is better than intuition? That's right. Math. She knows that she can use the angle and speed of her jump—a.k.a. her initial velocity—to estimate a landing position.

To do this, she starts by assuming she will be traveling in a straight line for a short amount of time. This gives her an estimate for where she'll be *if* she was, indeed, jumping along a straight line.

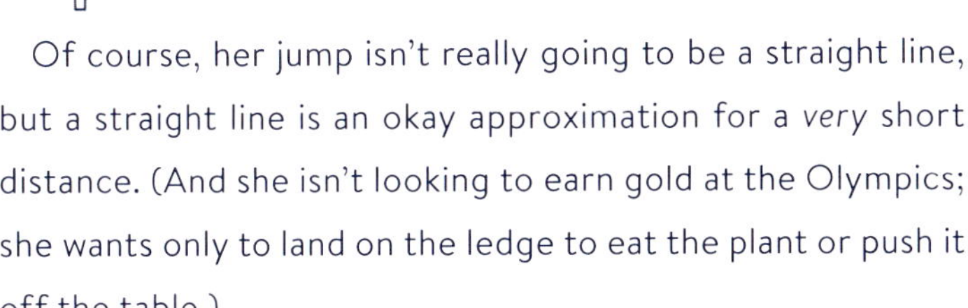

Of course, her jump isn't really going to be a straight line, but a straight line is an okay approximation for a *very* short distance. (And she isn't looking to earn gold at the Olympics; she wants only to land on the ledge to eat the plant or push it off the table.)

Now she knows a few things. First, where (approximately) she'll be after traveling that first short distance.

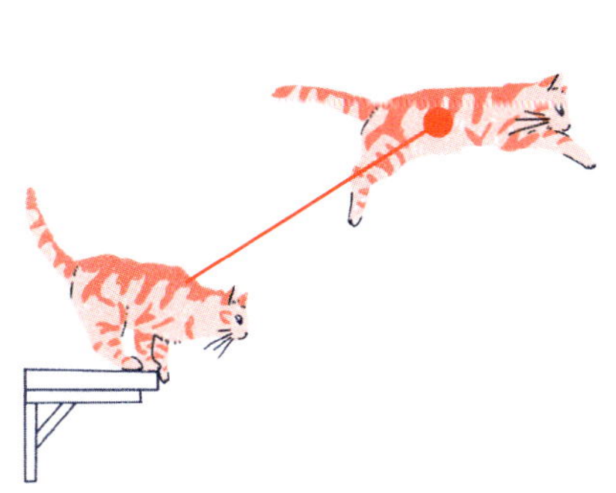

Second, she knows math. Using math and her predicted position, she figures out where she'd travel if, again, she flew along a straight line for a small distance, giving her another line segment attached to the first.

Which gives her the next predicted position along her jump.

She continues until finding where she'll land with her given initial velocity; that is, her given jump.

The path is dependent on how Butterball launches herself from the counter. (In mathematical language, we would say that her path depends on her initial velocity.) In the above example, her predicted landing spot is exactly where she'd want it to be.

However, changing her initial jump changes the steepness and length of the first line, which changes all other lines, changing her predicted landing spot. Here is an example where the prediction does not look good.

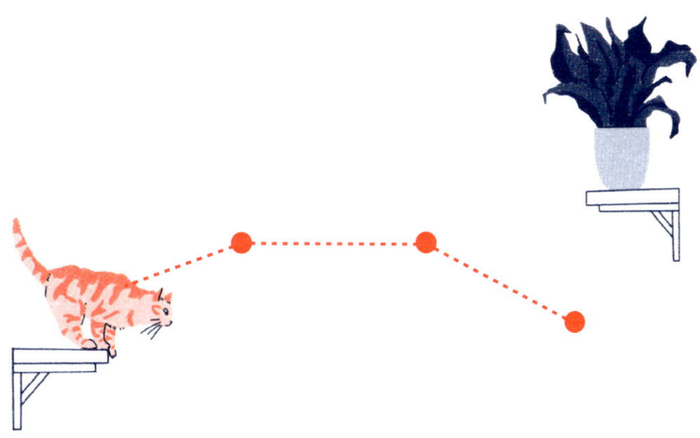

She can use this process to determine how she should jump to land at her desired spot.

Butterball is basically using a numerical method to approximate her flight path. Methods like this are used to approximate solutions to *differential equations*, which are equations relating something's rate of change to its value. These appear in calculus and are incredibly useful. For example, if we could derive a differential equation for how mercury levels in water are changing based on current conditions, we could determine when the water will be safe.

One difficulty with differential equations is that, even though they typically have solutions, most can't be solved by known methods. So, although we may know a solution exists, we often can't find it.

Luckily, even when you can't solve a differential equation, you may be able to approximate it, like Butterball. She assumed her velocity, which is the rate of change of her position, was constant for a short distance, predicted the value after that distance, then figured out her new velocity, and repeated the process. In essence, she approximated the path of her jump with a series of line segments, which is what happens in the first numerical method for differential equations that most students see: *Euler's Method*.

Numerical methods can produce astonishingly accurate approximations. Even though they are approximations (since, you know, Butterball is not really flying in little line segments) and will have error, that error can be decreased by taking shorter line segments. If the line segments are short enough, the error can be made small enough for the approximation to be of practical use. However, shorter line segments mean more line segments, requiring more calculations. (But it shouldn't be surprising that the quality of the prediction improves if you put in more work.)

Butterball, unfortunately, was impatient and used line segments that were too long. While her predicted path looked perfect, the error was large, and she missed the ledge.

Numerical methods like Euler's Method revolutionized the study of differential equations. Leonhard Euler (1707–1783) proposed this method in his book *Institutionum calculi integralis* (published 1768–1770). Euler was a prolific mathematician. As one small example of his contributions, he's the guy who decided we should use the Greek letter π to denote the mathematical constant π = 3.1415 . . . Euler was so intensely into his work that, upon losing the use of his right eye in 1738, he is quoted as saying, "Now I will have less distraction." Damn.

DOES A CAT EATING A BUTTERFLY IN KANSAS PREVENT TORNADOS?

Chaos Theory

You may have heard the saying "If a butterfly flaps its wings, it can cause a hurricane halfway around the world." But...what if a cat eats the butterfly? Does that have any effect? It turns out it could; if a butterfly flapping its wings can cause a hurricane, imagine how many hurricanes that cat just prevented?

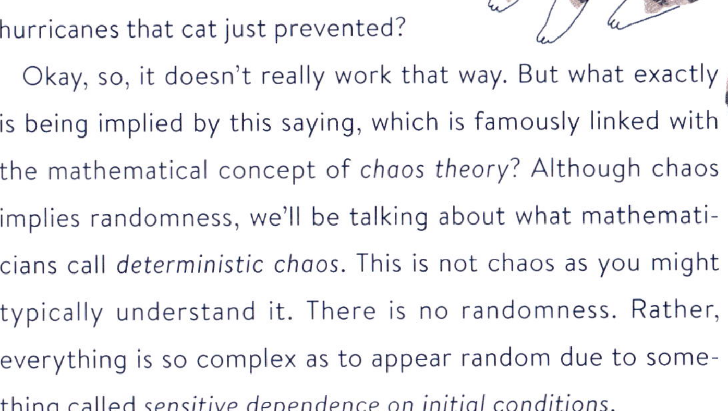

Okay, so, it doesn't really work that way. But what exactly is being implied by this saying, which is famously linked with the mathematical concept of *chaos theory*? Although chaos implies randomness, we'll be talking about what mathematicians call *deterministic chaos*. This is not chaos as you might typically understand it. There is no randomness. Rather, everything is so complex as to appear random due to something called *sensitive dependence on initial conditions*.

We need to unpack these definitions. When we can predict future outcomes exactly, we say we have a *deterministic system*. For example, cat researchers have shown that the number of cuddles a cat is willing to give depends precisely on the ounces of food they get each day. Butterworth, being a good kitty, has a fixed CPO (cuddles-per-ounce) of *exactly* thirty-six, meaning that for each ounce of food Butterworth gets per day, she will give thirty-six cuddles. And this ratio holds for all values; if she gets a half-ounce, you can expect eighteen cuddles.

So if you know *exactly* how much you fed Butterworth, you can determine *exactly* how many cuddles she'll give.

And, unsurprisingly, even if you don't know exactly how many ounces you gave Butterworth, as long as you are close, you can expect the predicted number of cuddles to be reasonably close as well.

So exact inputs give exact predictions in a deterministic system. The inputs are the initial conditions. Sensitivity to initial conditions means that the system is sensitive to small changes in the input. While not knowing exactly how many ounces you fed Butterworth wouldn't radically change the number of cuddles that day, if we have sensitive dependence, then a small change in input can lead to significantly different outputs.

Take Conan T. Cat, for example. Although Butterworth's CPO was deterministic and not sensitive to initial conditions, the same can't be said for Conan T. Cat. Although we can precisely determine how many cuddles we will receive if we know exactly how many ounces of food he has, it is sensitive to initial conditions. If food is measured out guaranteeing twenty-four cuddles, the slightest error could change that value to two cuddles or fifty-five cuddles. His cuddles depend sensitively on initial conditions, which isn't surprising, as Conan is a sensitive little guy.

We see sensitive dependence in natural settings beyond CPO. Consider two rivers. One is calm and the other turbulent. Assume that if we know the exact location of a ball floating in either river, we can predict precisely where it will be as it moves downstream. (This is a *big* assumption.) Unrealistically, we'll also say the rivers don't change over time, so that every time you throw a ball at a specific spot, it will always follow the same path. And this is true for any ball thrown at any spot, meaning the path of a ball based on where it is placed in the rivers is deterministic.

However, what if we throw two balls and they land close to each other? Will they stay close? In the calm river, we expect the balls to remain relatively close. But it is almost impossible to predict how close the balls will be in the turbulent river. They could wrap around a rock on different sides or get caught in eddies putting them far apart, or they could end up close.

So even if we can predict future locations exactly, that does not mean that inputs starting close to each other give outputs that are close. Approximation, which we depend on in applied mathematics, becomes the enemy. The smallest change can throw your predictions. For this reason, it is important to know when a system has sensitive dependence. Though there are a couple of conditions required for a system to be (mathematically) chaotic, sensitive dependence is the main condition that causes things to appear random.

The most famous chaotic system is probably weather. An early pioneer of chaos theory was Massachusetts Institute of Technology meteorologist Edward Lorenz (1917–2008). Using a computer model to predict a simple weather system, Lorenz and his collaborator wanted to see a sequence of data again, so they reentered the starting value as shown on a computer printout. However, they did not see the same result. Why? Maybe Lorenz's cat walked across the keyboard?

It actually wasn't an error! The computer printout displayed only three digits, while the computer itself was using six digits. A difference in those last three digits (something like 0.783 versus 0.783021) was enough to change the simulated weather entirely. The difference was small, and it was generally accepted that such a small change should have only small effects, but this was not the case. His model depended sensitively on initial conditions.

Lorenz's computer experiment is an important moment in the mathematical study of chaos theory. Weather was eventually shown to be chaotic, which guarantees that weather depends sensitively on initial conditions. So, for example, if we use temperature as one of the variables to predict the weather, then entering a current temperature of 59.08 degrees when it is really 59.07 degrees could make our prediction wildly inaccurate.

So, when the forecast is wrong, blame math.

Okay, what about those butterflies? If super-small changes in weather conditions can lead to large changes in the future weather, then the small changes to the wind patterns caused by a butterfly flapping its wings could, theoretically, cause a hurricane.

As mentioned earlier, chaotic systems always depend sensitively on initial conditions, but this is not all that makes a system chaotic. There are nonchaotic systems that also have sensitive dependence. (Similarly, cuteness is essential for a cat, but an animal being cute doesn't mean it is a cat.)

Interestingly, while weather is a chaotic phenomenon, climate—being the global average of weather conditions—is not. This relates to why we can't predict the weather a month from now, but we can predict global temperatures decades in the future, provided there are no unforeseen external changes.

The idea that a small change can cause a large future change has been around for a long time. One could argue that it is succinctly demonstrated in the proverb "For Want of a Nail," a version of which can be found in Benjamin Franklin's *Poor Richard's Almanack* (1758). Possibly the earliest version of this proverb was stated by the Middle High German poet Freidank in his collection *Bescheidenheit* (ca. 1230). This variant translates to "The wise tell us that a nail keeps a shoe, a shoe a horse, a horse a man, a man a castle, that can fight." Sensitive dependence also appears in Ray Bradbury's short story "A Sound of Thunder" (1952). Here, a time traveler accidentally kills a butterfly, leading to drastic changes when they return to the present. It is now difficult to find a time-travel story that does *not* involve some variation of this concept.

THANKS FOR ALL THE FISH

Infinity

Imagine you have a lot of kittens. And I mean a *lot* of kittens. Dreams come true, right? Every day you let them out into the large, but safely contained, backyard. Since not all cats go outside every day, it is important that you keep count. Counting so many kittens, while adorably fun, takes a long time. One simple solution involves a bunch of cat toys and a bucket.

For each kitten that goes into the yard in the morning, put a toy in the bucket. At the end of the day, remove a toy for every cat who returns home. If you are left with toys in the bucket, then not all cats have returned. Although you haven't counted, you know that the number of toys in the bucket is the same as the number of cats in the backyard.

This is an example of what mathematicians call a *one-to-one correspondence* or, if you want to be fancy like Fancy Feast, a *bijection*. These are used for, among other things, determining when two sets are the same size.

Here, a *set* is a collection of things called *elements*. A set could be a bunch of numbers, the chairs in your living room, the kittens who allow you to live with them, and so on. If two sets have a one-to-one correspondence between them, we say they have the same *cardinality*, meaning they have the same number of elements. Each element in either set has a unique "buddy" in the other set. Like the set of cat toys in the bucket and the set of kittens in the backyard.

What do we gain by doing a one-to-one correspondence rather than just counting the elements of each set? If we extend this idea to sets with infinitely many elements, then some really rad stuff happens. We'll show that, in some ways, infinity + 1 = infinity.

To start, let's make sure we understand what this means. Saying a set has infinitely many elements, or that the cardinality of the set is infinite, indicates that it has more elements than any finite set. Let's say the finite set has eight items. Yes, there are more than eight in the infinite set. What about twelve? Yup. And 312,333,204,586? Yes, there are more than

312,333,204,586 things in the set. To be infinite, this must be true of any number of elements.

Okay, so what practical use does this have? Let's be mathematicians and not worry about that. Instead, let's continue being impractical. Imagine an infinite number of cats are coming to your house for a fish dinner. You buy the fish and number them 1, 2, 3, . . . on your infinitely long dinner table. Since you have an infinite number of fish, this numbering keeps going. When the kittens arrive, you have each one choose a fish. Cats aren't rude, so they wait, not wanting to eat until everyone is ready. You prepared well, and each kitten gets a fish and there are no fish left over.

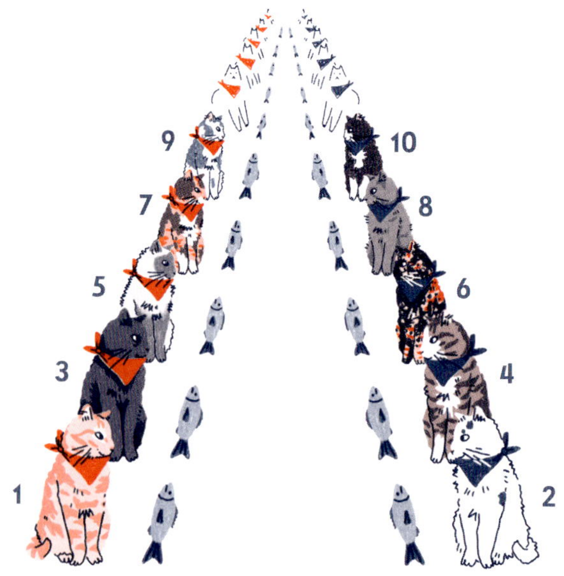

There is a one-to-one correspondence between the set of kittens and the set of fish; those sets are the same size. That seems obvious, right? Because I said there were an infinite number of each?

The kittens are just getting ready to eat when Pudding Cup, who is always fashionably late, arrives, well... fashionably late. Too bad for Pudding Cup, right? All the fish are spoken for. If each fish has a kitten and each kitten has a fish, there is no fish left over for Pudding Cup. That would be true if we had finitely many kittens and fish, but both sets are infinite, and infinity is weird.

You can make room for Pudding Cup by asking the kitten at fish number 1 to move to fish number 2. And the kitten at fish number 2 to move to fish number 3 while the kitten at fish number 3 moves to fish number 4. And so on. Every kitten moves to the next fish. We can do this because for any numbered fish we know that there is another after it because the set is infinite. Since each cat can move to a new fish, then every cat who arrived on time still has a fish to themselves. It is just numbered one higher.

But notice that this leaves fish number 1 unspoken for! Pudding Cup can slide right in! Now that everyone is here, the kittens can eat!

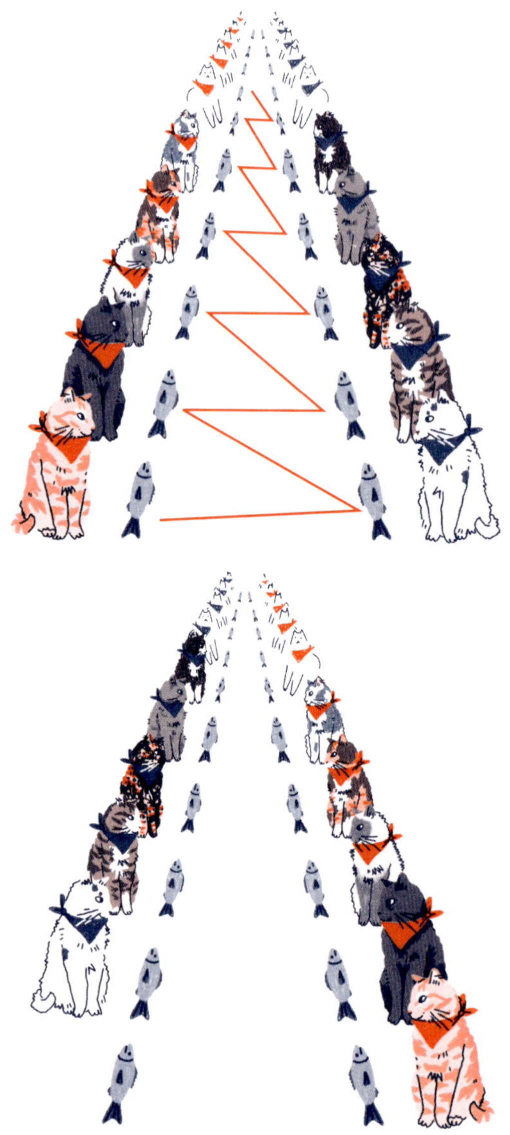

The set of cats who were on time and the new set of those cats plus Pudding Cup were each put in one-to-one correspondence with the set of fish. Since the set of all cats is just the set of cats who were on time plus one additional cat, we see that, in this sense, infinity + 1 = infinity!

So far, we've ignored applications, but this is very useful when you are arguing with your cat about which of you loves the other more. Once you hit the "I love you infinity" phase, you know that "I love you infinity + 1" doesn't advance the game.

A TALE OF TWO KITTIES

Amicable and Perfect Numbers

Numbers can carry hidden meanings. For example, did you know that math can tell you why six is a perfect (or should that be purrfect?) number of cats to own? Or that the pendants worn by Milo and Buttons tell us they are best friends?

It all comes down to *factors*. For a natural number (also called counting numbers, like 1, 2, 3, ...), a *factor* is another number that divides evenly, without remainders, into the first. So, two is a factor of ten because 10 ÷ 2 = 5. Note that every number has one and itself as factors.

If a number has *only* these two factors, the number is prime, while others are composite. The number one, however, is a bit of an oddball and is neither prime nor composite. We require that prime numbers have two *distinct* factors. Loneliest number, indeed.

The Pythagoreans, the philosophical sect (or, as some believe, cult) following the teachings of Pythagoras (from page 2), were very interested in the mystical properties of numbers. They looked for patterns *everywhere* in numbers. In particular, they were interested in *perfect numbers*, and six is one of them.

What's so special about six? Consider the *proper factors* of six. These are the factors of six, except for six itself. The factors of six are 1, 2, 3, 6, while the proper factors are 1, 2, 3. Now let's add the proper factors: 1 + 2 + 3 = 6. Whoa! We get six back. This is not usually true; for example, for eight we get 1 + 2 + 4 = 7. Close, but not equal. A number is called *perfect*... sorry, I mean *purrfect*, if it is the sum of its proper factors.

The next purrfect number is 28 = 1 + 2 + 4 + 7 + 14. After that is 496, and then 8,128. Given that we have infinitely many numbers to choose from, it may seem that there should be infinitely many perfect numbers. Maybe, but right now we know only of 52 perfect numbers, with the 52nd discovered in 2024. There is no proof that the collection of perfect numbers is infinite and no proof that it is finite. (Sounds like a problem for Schrödinger's cat if he were into math.) Also, we have never found an odd perfect number. They are either rare or nonexistent.

While six is *a* purrfect number of cats, it is not *the* purrfect number of cats since you have other purrfect numbers to consider. But be warned, finding an apartment for you and your 28 (or 496) cats could be tough.

Okay, so what's the deal with the pendants? Being perfect was all about adding your proper factors. But sometimes numbers, while not perfect on their own, are kinda perfect through friendship. Consider the numbers 220 and 284, as worn by Milo and Buttons.

NUMBER	PROPER FACTORS	SUM OF PROPER FACTORS
220	1, 2, 4, 5, 10, 11, 20, 22, 44, 55, 110	284
284	1, 2, 4, 71, 142	220

Definitely not purrfect, but the sum of the proper factors for each number adds to the other number! Pairs of numbers like this are called *amicable numbers*.

Other pairs of amicable numbers include 1,184 and 1,210, 2,620 and 2,924, and 5,020 and 5,564. We have no idea whether there are infinitely many pairs of amicable numbers, although we know of more than one billion such pairs.

Naturally, the Pythagoreans knew about amicable numbers and associated mystical properties to such pairs. In particular, 220 and 284 were used as symbols of friendship, which is why Milo bought these pendants. Buttons needed convincing that anything about math could be amicable. He was not a strong student.

COUNTING BASE-MEW

Binary

Here are 10 cats:

Wait...ten cats? Are eight hiding? Is this wishful thinking? Am I hitting the catnip too hard? Turns out, none of the above. Rather, I'm counting in binary.

Binary? Isn't that the strings of 0s and 1s that let sci-fi robots talk to each other? Well, yes, but it is also a pretty nifty mathematical concept.

You may be familiar with the names for positions in a number. Namely, the "ones" place, the "tens" place, the "hundreds" place, and so on. We use these names because we write numbers as sums of powers of ten—a power of ten is just a fancy way of saying a bunch of tens multiplied together. For example, when we write 1,011 we mean 1,000 + 10 + 1, or the sum of one 1,000, no 100s, one 10, and one 1. (Hence the name for the places as ones, tens, hundreds, and so forth.) The use of powers of ten (for example, 1,000 = 10 × 10 × 10 = 10^3) is what makes it base-10.

For most people, every number they encounter is written base-10. But there are other numerical bases, one of which is base-2, a.k.a. binary. If base-10 is all about powers of ten, then base-2 is, you guessed it, all about powers of two.

What effect will this have? Well, instead of a thousands place, where 1,000 = 10 × 10 × 10, we'll have an eights place, since 8 = 2 × 2 × 2. We are just switching tens for twos. This means that the digits making up 1011 have a different meaning in base-10 than they do in base-2.

Converting from binary to base-10 is actually pretty simple. In fact, it is the same for any base conversion: we just add up the powers. Our good friend 1,011 base-10 means:

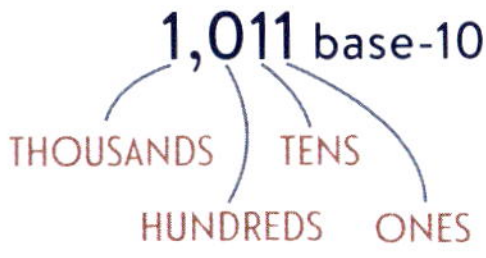

$$1{,}011 \text{ base-10} = 1 \times 1000 + 0 \times 100 + 1 \times 10 + 1 \times 1$$
$$= 1 \times 10^3 + 0 \times 10^2 + 1 \times 10^1 + 1 \times 10^0$$

THOUSANDS HUNDREDS TENS ONES

But what about 1,011 base-2? We use the same process we used for base-10, but using twos instead of tens. We then add up what we get.

$$1{,}011 \text{ base-2} = 1 \times 2^3 + 0 \times 2^2 + 1 \times 2^1 + 1 \times 2^0$$
$$= 1 \times 8 + 0 \times 4 + 1 \times 2 + 1 \times 1$$
$$= 8 + 0 + 2 + 1 = 11 \text{ base-10}$$

EIGHTS, FOURS, TWOS, ONES

So, if you are speaking base-10, you would say eleven, but in base-2 it is 1,011. It's kind of like translating between languages. If you have a pile of apples in front of you then the word you use for the number of apples depends on the language you are speaking. Similarly, someone using a different base will use different digits. The numbers won't look the same, but they both represent the same quantity.

What about our claim of ten cats? Well, ten base-2 means we have one 2 and no 1s, meaning the number is two base-10. So having two cats (base-10) is the same as having ten cats (base-2). *This is the only time I will ever say that ten cats are not better than two.*

Why would we care about binary/base-2? Well, besides being the base (pun definitely intended) for a lot of nerd jokes, binary is at the heart of computer languages. Here, binary uses 0s and 1s to represent whether various electrical switches are "on" or "off," with a 1 being "on" and a 0 being "off." In fact, the International Electrotechnical Commission standard symbols for "on" and "off" look like 1s and 0s.

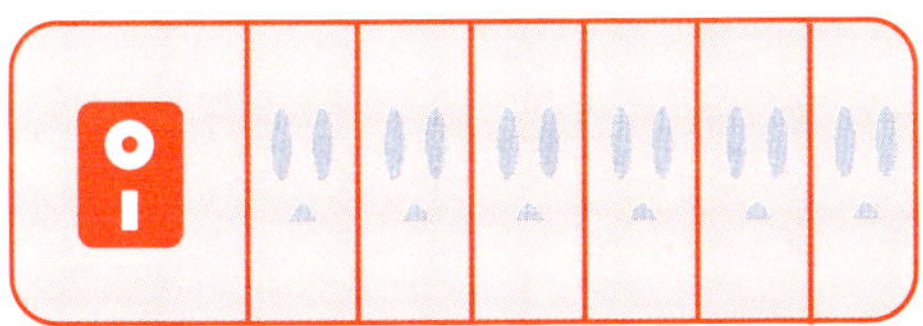

And the universal symbol for a power button is a combination of these two symbols.

Okay, enough about binary...time to give my cat a belly scritch. As they say, my cats may have 1,100,011 problems, but a belly scritch ain't one.

The numbering system we are most familiar with is base-10. There are a few ways to see why it is called this. For one, there are ten unique symbols: 0, 1, 2, . . . , 9. These are combined in various ways to represent numbers. The position matters, which is why this system is sometimes referred to as a *base-10 positional numbering system*. (If you want to sound fancy, grab a cane and a monocle, and then drop that phrase at your next cocktails-and-cats party.) In binary, or base-2, we have exactly two symbols: 0 and 1. For any base, the number of symbols will match the base.

LOAFING AROUND

Topology

Cats love to loaf around... by which I mean lying around looking like loaves of bread.

Cats have the amazing ability to change their shape, fitting into small boxes and bowls where they have absolutely no right fitting.

To a mathematician, this is just elementary topology, an extension of geometry where objects don't need to be rigid and can be stretched, bent, or compressed.

Imagine, for example, that geometric shapes are made of Silly Putty. Anything you can shape from Silly Putty is considered topologically equivalent to what you started with as long as you do not glue (that is, stick two pieces together), cut, or puncture the putty. You can bend and stretch all you like, however.

As an example, a cat made of putty is topologically equivalent to a loaf of bread since we can reshape the kitten into a loaf of bread without cutting, gluing, or puncturing.

However, even though cats curl up into little donut-like shapes, they are not topologically equivalent to a donut. Their tail and mouth may be touching, giving the illusion of a donut-shaped object, but they are not attached. Attaching the tail and mouth would break the rule against gluing. And it is just rude.

In 1858, mathematicians Johann Benedict Listing (1808–1882) and August Ferdinand Möbius (1790–1868) independently discovered a fascinating mathematical object now known as the *Möbius strip*. To create the Möbius strip, start with a strip of paper, give it a half-twist, and attach the two ends together.

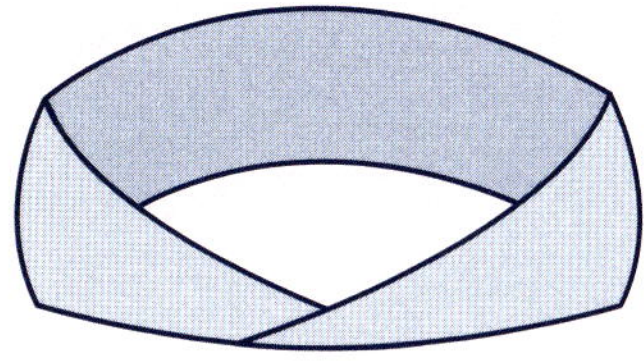

This topological curiosity is remarkable for a few reasons. First, it has only one side. This may sound impossible, but if you start on one side and trace around the strip, you end up where you started . . . but on the other "side" of the paper. Continue tracing, and you find yourself back at the starting point! Conveyer belts have even been designed with a half-twist to ensure equal wear on both sides. (Your car's fan belt could be a Möbius strip!)

Although the Möbius strip was discovered mathematically in 1858, Roman art from the third century BCE depicts its form. And, of course, cats were entertaining kittens with the amazing properties of the Meöwbius strip long before that.

This may seem like mathematical nonsense, which is okay since that has never stopped a mathematician from mathing, but there are applications. As an example, Scraps is trying to get to the Museum of Feline Arts near Tortie Street in Kitty City to see the latest Salvador Meowli exhibit. Unfortunately, this is the subway map for Kitty City:

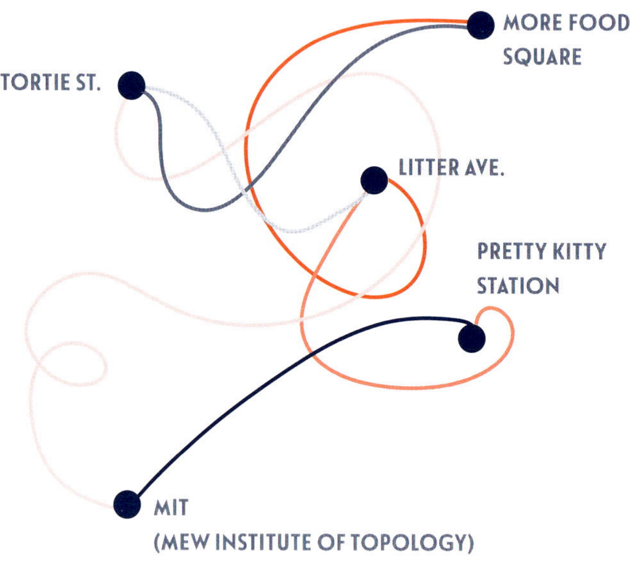

As you can see, the engineers hired by Kitty City were not great at planning infrastructure. And the map is confusing. It is hard for the cats riding the rails to know where their next stop is, and what does it mean when two subway lines cross, but the intersection is not labeled as a stop? Can you still switch trains? How would Scraps ever reach the exhibit using this as a map?

Luckily, Professor Bojangles teaches at MIT (the Mew Institute of Topology) and knows that the stops are the most important information, not all the twists and turns. Treating the map topologically, the good prof straightens the curves representing the subway lines and obtains the following map:

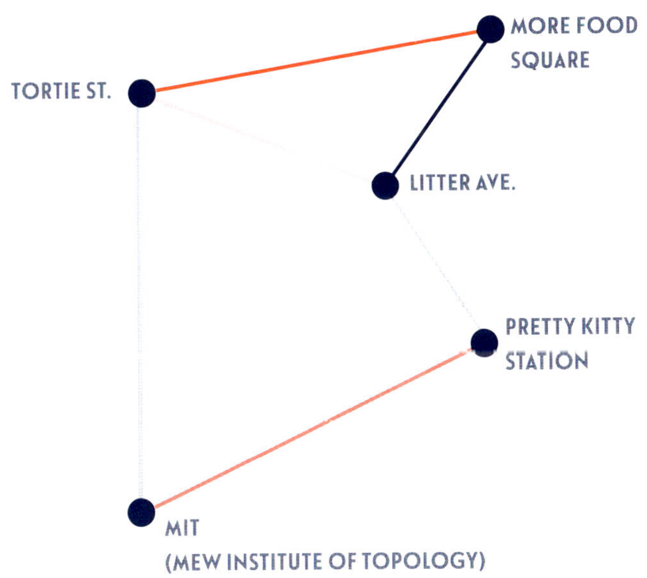

With this map, Scraps has less trouble finding the exhibit than he does sleeping in a sunbeam.

Many cities use subway and train maps that are not geometrically accurate (straight routes on a map do not mean the train is moving in a straight line) but are topologically equivalent to the true map.

Beyond maps, topology is used by engineers to determine structural stability and by computer scientists to increase the efficiency of computer networks.

And, of course, cats use it daily to fit in cute places, making us go "awwww!!!!" and getting clicks on social media, thus creating jobs.

MR. PEPPA'S PARTY PROBLEM

Birthday Paradox

Cats like to party, and the most common party is in honor of a cat's cat-day.

A cat-day is kind of like a birthday. When it is a cat's cat-day, there is usually a huge cat-day party thrown with a lot of cake and dancing. This year, Mr. Peppa is in charge of organizing the parties for his block. Mr. Peppa hopes no two cats on his block have the same cat-day, as that requires extra prepping to ensure everyone has a good time. (And cats can be divas; they do not like sharing cat-day parties, although they will if they need to.)

Mr. Peppa's in luck, though, as there are only thirty-two cats on his block! So, it should be pretty unlikely that two share a cat-day, right?

This is a job for the *Cat-Day Problem*, which asks for the probability that if you have a bunch of randomly chosen cats, at least two share a cat-day. (A related problem is the *Birthday Paradox* for human birthdays.)

With 365 possible cat-days, we know that if there were 366 cats on Mr. Peppa's block, there would have to be a shared cat-day. But what about thirty-two cats? Surprisingly, if you have thirty-two cats, the probability that two of them share a cat-day is an astounding 75.33%! Uh-oh, Mr. Peppa!

So where does 75.33% come from? Let's approach the problem in reverse and find the probability that we do *not* have any shared cat-days. Pick a cat at random and ask for their cat-day. Then approach a different cat. The probability that the second cat has a different cat-day than the first is $364/365$ because there are 364 days that currently do not have a cat-day. Assuming we have two cats with different cat-days, the probability that a third cat does not share *either* of the earlier cat-days is $363/365$. Why 363? There are now two days that are already taken with cat-days. This means that the probability that none of the three cats shares a cat-day is $(364/365) \times (363/365) \approx 0.9917$, or 99.17%.

If we want the probability for thirty-two, we use the same idea as we did for three cats. This time we get

$$(^{364}/_{365}) \times (^{363}/_{365}) \times (^{362}/_{365}) \times \ldots \times (^{334}/_{365}) \approx 0.2467$$

This means there is a 24.67% chance that we do *not* have a shared cat-day, giving us a 100 − 24.67 = 75.33% chance that we *do* have a shared cat-day! Whoa!

In fact, we need only twenty-three cats to give us a greater than 50% chance of a shared cat-day! Welcome to the *Cat-Day Paradox*. How can only twenty-three cats give a greater than 50% chance of a shared cat-day? I mean, there are 365 possible cat-days and only twenty-three cats!

We may have only twenty-three cats, but since we are interested in shared cat-days, we really want to think about how many pairings of cats there are. For example, how many unique pairs of cats can we make from the trio of Butters, Tofu, and Pop-Tart?

We can choose any of the three cats first, leaving two cats. This means forming a pair can be thought of as a choice of three followed by a choice of two, giving 3 × 2 = 6 total options.

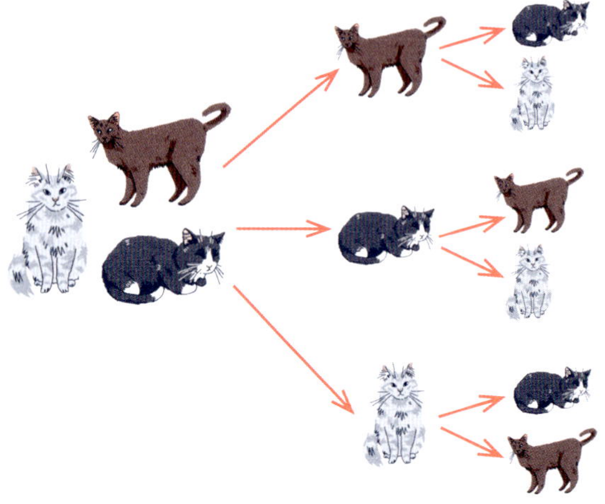

But we are accidentally double counting. Choosing Butters and then Pop-Tart gives the same pairing as choosing Pop-Tart and then Butters. We need to divide our count by two, giving us 6 ÷ 2 = 3 pairings (see opposite page).

If we have twenty-three cats, then there are twenty-three choices for the first cat and twenty-two for the second, giving us 23 × 22 initial options, but we again need to divide by two since order doesn't matter. This means we have $\frac{23 \times 22}{2}$ = 253 pairs of cats to check for shared cat-days, which is more than half the days in a year.

Historically, it is thought that English mathematician Harold Davinport (1907–1969) originally stated the Birthday Problem in 1927, but never published it because he "could not believe that it had not been stated earlier." The problem was first published in 1937 by Austrian mathematician Richard von Meses (1883–1953).

Related to the Cat-Day Problem is the *Cat Collector's Problem*. Imagine there are ten breeds of cats available at a shelter. (For this to work, the shelter must be very large with lots and lots of each breed.) You want one of each breed. The problem—you can't see the cat you adopt until after you've adopted. How many cats would you need to adopt, on average, to ensure you have at least one of each breed?

I have good news and bad news. (Or should that be good mews and bad mews?) The bad news is that you need to adopt thirty cats to expect one of each of the ten types. The good news is that you need to adopt thirty cats.

But why thirty?

Once you have a certain number of breeds of cat, we can find the probability that the next cat adopted is not a breed you have. For example, if you currently have eight different breeds, there are only two left, so you have a $2/10 = 1/5$ chance of getting a new breed on your next adoption. We can use this to get the expected number of adoptions needed to get the next breed. With a $1/5$ chance of getting a new breed, we expect it to take five adoptions to get our next new breed.

What if the math isn't so easy? What if we have only one breed? Then we have a $9/10$ chance of a new breed on the next adoption. Since $9/10$ is the same as $1/1.111$ (with rounding), we expect it to take 1.111 adoptions to get the second breed.

Where did 1.111 come from? You may remember this from algebra, but if we want to write $9/10$ as a fraction with a one on top, we can use the fact that $9/10 = \frac{1}{10/9}$ and $10/9 \approx 1.111$. Hence, a $1/1.111$ chance.

Okay, let's run with this. If we have a $9/10$ chance of getting a new breed, we expect it to take $10/9$ adoptions. And, I know, we can't do fractional adoptions, so what does this even mean? If we had thousands and thousands of people trying to get a second breed and we asked them all how many adoptions it took, we'd get a lot of ones, some twos, a few threes, but if we averaged all those values, we'd expect around 1.111.

Now that we can go from probability to number of expected adoptions required, we can make a table to find out how many adoptions are needed in total.

NUMBER OF UNIQUE BREEDS OWNED	PROBABILITY OF NEW BREED	NUMBER OF ADOPTIONS TO GET A NEW BREED	NUMBER OF ADOPTIONS TOTAL
0	1	1	1
1	$^9/_{10}$	$^{10}/_9 \approx 1.111$	2.111
2	$^8/_{10}$	$^{10}/_8 = 1.25$	3.361
3	$^7/_{10}$	$^{10}/_7 \approx 1.429$	4.79
4	$^6/_{10}$	$^{10}/_6 \approx 1.667$	6.457
5	$^5/_{10}$	$^{10}/_5 = 2$	8.457
6	$^4/_{10}$	$^{10}/_4 = 2.5$	10.957
7	$^3/_{10}$	$^{10}/_3 \approx 3.333$	14.29
8	$^2/_{10}$	$^{10}/_2 = 5$	19.29
9	$^1/_{10}$	$^{10}/_1 = 10$	29.29

We expect that it'll take roughly 29.29 adoptions to collect all ten breeds. Since an individual person can't do 29.29 adoptions, they need to do thirty.

This is also called the *Coupon Collector's Problem* because it is used to calculate how many items you can expect to purchase to win in a "collect all coupons and win" game.

You can relate the Cat-Day Problem and the Cat Collector's Problem by thinking of how many cats you need in a room to ensure you have a cat-day every day of the year. Imagine the table above, except going from 0 to 364 instead of 0 to 9. Since I'd rather pet a cat than make that table, I'll just let you know that it takes around 2,365 adoptions.

But think about it. If you have a cat-day every day, then every day is a party! This seems like an appropriate reason to have 2,365 cats, if you ask me. The math gets harder, however, if one of those cats is Puddles, whose cat-day is a leap day...

KEYBOARD CAT

The Infinite Monkey Theorem

Mrs. Waffles likes to walk across keyboards, typing gibberish, especially while her roommate is trying to work. If she does this long enough, will she eventually write out "LUV YOU"?

Maybe, but it depends on some assumptions for how the cat is walking.

To make things easier, let's use the following keyboard. (Since Mrs. Waffles likes walking across keyboards so much, we got her a custom one of her own.)

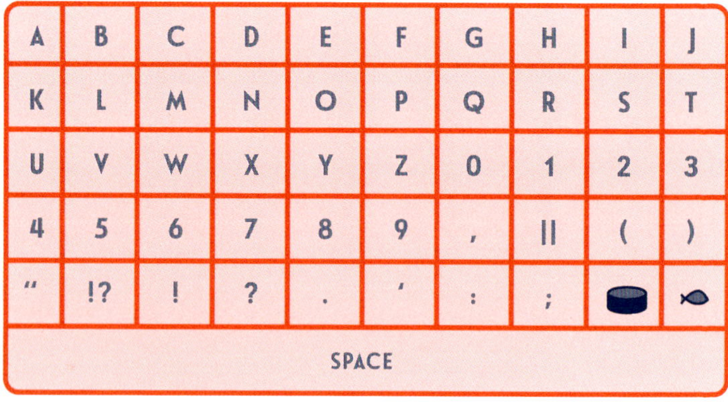

If you ignore capitalization, you can write most things with this keyboard. But will Mrs. Waffles write "LUV YOU"? You may have noticed the two cat-specific keys. The cat-food button dispenses food for the cat, and the toy button gives them a cat toy when pressed.

Given that, I'm not sure I expect Mrs. Waffles to write anything. More than likely she'll figure out walking around the area of the keyboard with the food and toy keys leads to rewards, so she'll write very little and most likely nothing that will form words.

Even ignoring those keys, a cat walking around is more likely to hit certain key combinations than others. If Mrs. Waffles hits the *T* key, it is more likely she'll hit *R* next rather than the much farther away *U*.

Let's refine the problem. Assume that Mrs. Waffles hits all keys with equal probability, *and* her key presses are independent of each other (meaning, for example, even if she hits a *T*, both *U* and *R* are equally likely for the next key).

In this case, since all keys are equally likely, there is a $1/50$ chance Mrs. Waffles steps on a particular key. "LUV YOU" takes seven characters, counting the six letters and the press of the space bar. Let's further assume that when Mrs. Waffles walks across a keyboard, she always hits seven keys. We'll refer to Mrs. Waffles typing seven characters as a *stroll*.

What is the probability that Mrs. Waffles types "LUV YOU" on a single stroll? The chance she gets an *L* first is $1/50$.

If she wants an *L* and then a *U*, she has a $(1/50) \times (1/50) = 1/50^2$ probability. The probability she writes "LUV YOU" during one stroll is $1/50^7$, or $1/781{,}250{,}000{,}000$. So around 1 in 780 billion or 0.000000000128%. This is insanely small, but it is not zero.

For comparison, a 1 in 302 million chance of winning is not uncommon for large-payout lotteries, meaning you are roughly 2,582 times more likely to win millions playing the lottery than you are to see "LUV YOU" written by Mrs. Waffles in a single stroll. *Do not confuse this low probability with the probability that Mrs. Waffles loves you.*

However, even though the probability is small on a single stroll, it increases the more strolls she takes. In fact, as her strolls increase, the probability that one of them spells "LUV YOU" approaches one, meaning that we would expect "LUV YOU" to eventually show up. But how many strolls are we talking about? Mrs. Waffles has other things to do, you know.

Since a single stroll gives a $(1/50)^7$ chance of writing "LUV YOU," there is a $1 - (1/50)^7$ chance that she does *not* write it.

What about two strolls? Then she has a $1 - (1/50)^7$ chance of *not* writing "LUV YOU" for each of the two strolls, giving a $\left(1 - \left(\frac{1}{50}\right)^7\right)^2$ chance of neither stroll being "LUV YOU." This pattern continues: if she take fifty strolls, she has a $\left(1 - \left(\frac{1}{50}\right)^7\right)^{50}$ chance that none of the strolls say "LUV YOU."

The following table shows probabilities for various strolls.

STROLLS ACROSS KEYBOARD	PROBABILITY OF NOT GETTING "LUV YOU"	PROBABILITY OF GETTING "LUV YOU"
1	100%	0%
1,000,000,000	99.9%	0.1%
100,000,000,000	88%	12%
1,000,000,000,000	27.8%	72.2%

Notice that the probability of getting "LUV YOU" is approaching 100%...but wow does it take a long time. And this was only for one specific seven-character combination. If you want Mrs. Waffles to write the next great American novel, she'll definitely get her steps in. Luckily, if you are able to watch Mrs. Waffles at work, it is probably worth it!

Striking keys at random may remind you of the Infinite Monkey Theorem, which states that a monkey randomly hitting keys for an infinite amount of time will eventually type out all literature. By why talk monkeys when we have Mrs. Waffles?

THE FOUR SCARVES OF PURRIDISE ISLAND

The Four-Color Theorem

Purridise Island is a small island run by cats. The island is divided into twenty-one villages, with cats identifying their village in the cutest way possible: colored bandannas.

Since these cats don't travel overly far, they need to ensure only that the color of a village's bandanna is different from the colors of the bandannas for all villages bordering that village. Below is a map of Purridise Island and its twenty-one villages. Given the conditions above, how many different bandannas do we need?

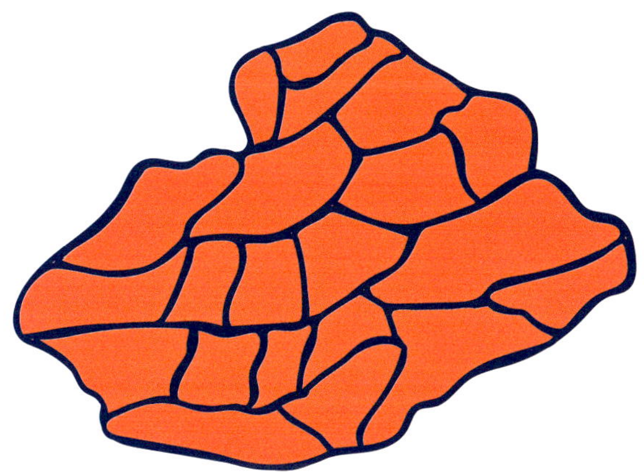

Since there are twenty-one villages, we would certainly be set with twenty-one colors...but we want the minimum number of colors.

It turns out, we need only four.

An example coloring is shown here:

Surely this depends on the number of villages and how they are arranged, right?

Wrong.

It turns out that four colors will always suffice. This result is known as the *four-color theorem*. Mathematicians believe that the first statement of this theorem was given in 1852 by Francis Guthrie (1831–1899) when he was coloring a map of the counties of England and noted that he needed only four colors. Of course, that could depend on the map he was using. He tried other maps...and it turned out that he was able to color any map using at most four colors.

So, is it true that for any map (with certain restrictions), no more than four colors are required? How does one prove this is true for *all* maps and not just the ones Guthrie happened to look at? For a mathematician, this is a tasty problem.

Guthrie first checked whether others had explored the hypothesis, bringing the question to mathematician Augustus De Morgan (1806–1871). Unfortunately, De Morgan didn't know.

The question then appeared in the British magazine the *Athenaeum* in 1854 by someone credited as "F. G.," which could be either Francis Guthrie or his brother Frederick Guthrie. No answer was given...and it turned out that this was a difficult problem to tackle. Over the following years, many famous mathematicians shot their shot. It was easy enough to construct maps where three colors were not enough, and no one could draw up a map where five colors were needed. But that doesn't mean such a map doesn't exist.

Proving something doesn't exist (like a map requiring at least five colors) is pretty hard. To demonstrate how hard this can be, consider the related *fur-color hypothesis*, which states that no cat has more than four naturally occurring fur colors. (It is a hypothesis because it has not been proven true, in which case it would become a theorem.)

If you can find a cat with five or more fur colors, then you have shown that the fur-color hypothesis is not true. But what if you can't find such a cat? Does that prove they can't exist? To prove they can't exist, you would need to show that something about being a cat prevents more than four colors.

In 1976, more than one hundred years after the original question was asked, mathematicians Kenneth Appel (1932–2013) and Wolfgang Haken (1928–2022) announced that they had proven the theorem. However, their proof was assisted by a computer program checking more than a thousand specific maps over the course of more than a thousand hours. This was the first example of a computer-assisted proof of a major mathematical theorem, and it caused quite a bit of controversy.

To further complicate matters, they submitted more than four hundred pages of microfiche with diagrams and more explanation. For those who don't know, microfiche sheets contain multiple regular pages shrunk and read with a magnifier. A typical sheet of microfiche can contain between forty and one hundred "regular" pages of print...so four hundred pages of microfiche is a lot. Appel and Haken rewrote the proof, simplifying and addressing some questions, in their 1989 book *Every Planar Map Is Four Colorable*.

The four-color theorem is one of those mathematical results that is easy to state but surprisingly difficult to prove. To this day, no proof has been offered that does not rely on computer assistance to check many individual cases.

Thankfully, we do not need to understand the proof to help the kittens from Purridise Island set up their ethically sourced bandanna factories.

HOW LONG IS THE COAT OF MITTENS?

The Coastline Paradox

Mittens wants to go to the high-end groomer Vidal Maine Coon. However, Vidal Maine Coon charges based on *cat-purrimeter*, which is the perimeter of a high-resolution image of the cat. Mittens has a meager allowance, so to see if he can afford the cost, Mittens needs his cat-purrimeter.

Figure 1: Insanely high definition photo of Mittens

Since Mittens isn't made of lines or circles, it can be hard to get the length "around" him. How should we approximate this? One method is to use lines of a fixed length to try to go around the outline of Mittens. (We'll most likely need a partial length for the final line segment.)

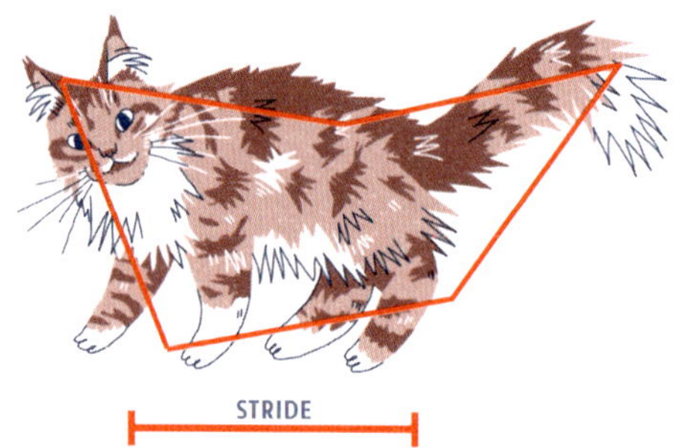

Imagine being small enough to walk around Mittens's image. (Do **not** walk on Mittens.) If we fixed our stride and connected our locations with lines, we'd get the above image. Therefore, we'll call the length of the line segments the *stride*.

If we take smaller strides, we'll get more lines that more closely approximate Mittens.

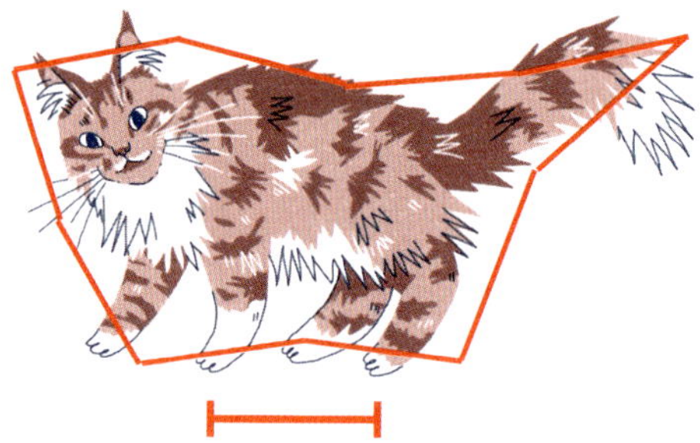

As your stride decreases, the cat-purrimeter estimates increase. Which makes sense... When stride decreases, we "hug the curves" more and skip less of Mittens. So how do we know when our stride is small enough to get an accurate enough estimate?

That can be tough to answer. In fact, a similar question about how stride affects distance birthed a new type of geometry. It wasn't that mathematicians were trying to calculate cat-purrimeters. Rather, they were trying to measure the length of a coast.

Around 1951, Lewis Fry Richardson (1881–1953) noticed that Portugal and Spain reported different lengths for their shared border, which includes several rivers. The difficulty with measuring coastlines and riverbanks is that they are crinkly, and crinkliness isn't great for measuring length. In fact, mathematicians noticed that as you decrease your stride, the length of a coast increases, and theoretically without bound! So, coasts are, in some ways, infinitely long?

To understand this, let's go back to Mittens. When using a stride that is smaller than the width of an individual hair, you need to go "around" the hair, adding distance.

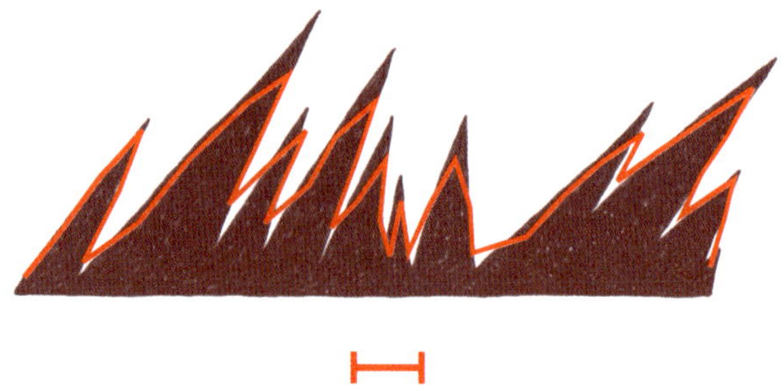

As the stride continues to decrease, you move around more and more hairs, giving more and more length. With coasts, the added length comes from moving around smaller rocks and moving farther up streams and rivers . . . When does the coast stop being the coast and become a riverbank?

The idea of coasts not having a well-defined length is the *Coastline Paradox*, noted by Hugo Steinhaus (1887–1972) in his 1951 paper "Length, Shape, and Area." Lewis Fry Richardson wrote about the paradox in "The Problem of Contiguity: An Appendix to Statistics of Deadly Quarrels" (banger of a title, Lewis), and Benoit Mandelbrot (1924–2010) continued, publishing the now famous "How Long Is the Coast of Britain? Statistical Self-Similarity and Fractional Dimension" in 1967.

Mandelbrot's explorations into the Coastline Paradox led to the development of *fractal geometry*. Fractals are insanely interesting (and often beautiful) geometric objects with a lot of strange properties. And, it turns out, fractal geometry is much better suited for describing objects in nature than the standard geometry of lines and circles. Fractals are well worth a dive down the rabbit hole. (Or is that a dive into the cat crate?)

Luckily, Mittens found that, buried in the fine print, Vidal Maine Coon specifies a stride, allowing him to figure out his purrimeter. He is extra excited because it is within his budget! Move over, Paul Mewman, there's a new handsome guy in town.

I HEARD THE MEWS TODAY...

Graph Theory

Sir Cuddles is starting a new job delivering mail and the *Daily Mews* for the hamlet of New Skritzersōhne.

Since this is his first day with the USPSpsps, he wants everything to go smoothly. He has mapped out his route, and it seems easy enough. Since he is a beginner, he has only a few stops, and, as luck would have it, he was assigned the district with New Skritzersōhne's famous bridges!

He really wants to walk each of the bridges each day. Is there a way for him to cross each of these bridges starting and ending at his house? Well, sure...that isn't hard, but he takes his job as a government employee seriously, so efficiency is important. He wants not only to cross each bridge but also to cross each bridge *exactly* one time. That's a more difficult mouse to catch.

Math has good news and bad news for Cuddles. The good news? Math knows whether this is possible. The bad news? It isn't. But more good news? There is a whole branch of math, *graph theory*, that looks at questions like this, which can help Sir Cuddles should he get to pick his next route.

You may have nightmares about graphs from algebra and calculus class, but the graphs in graph theory are different. Here, we have a bunch of dots, called *vertices*, connected by a bunch of lines, called *edges*.

We make a graph from New Skritzersöhne by replacing landmasses with vertices and bridges with edges. The edge lengths and shapes don't matter, only which vertices they connect. This is similar to Professor Bojangles's approach to Kitty City's subway map on page 65.

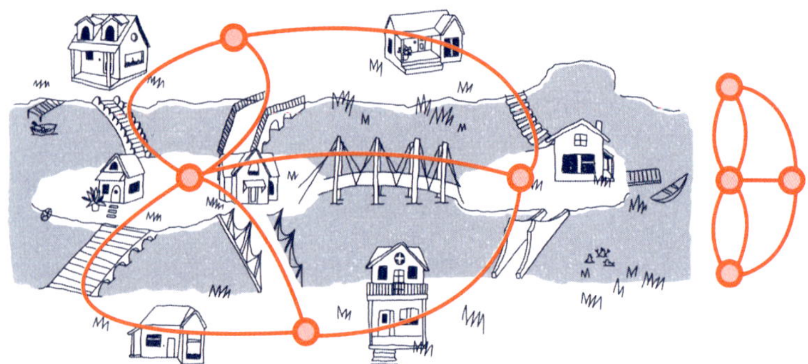

Cuddles's question can now be stated as whether he can start at a vertex and walk along each edge exactly one time, ending where he started. Cuddles wants an *Eulerian circuit*, which is a path around a graph, starting and stopping at the same vertex and crossing each edge exactly once.

Eulerian circuits are named for Leonhard Euler (for more about Euler, see "Euler's Method" on page 32), who tackled the same problem, with the same graph, as Cuddles in 1736. The layout of New Skritzersöhne, as a graph, matches that of the Prussian city of Königsberg (now Kaliningrad, Russia). The story goes that each Sunday, the citizens would walk about Königsberg. As a fun side quest, they tried to find a way to walk across each of Königsberg's seven bridges exactly once, ending where they started.

Unfortunately for Cuddles, Euler showed that such a route is not possible. We can see this by considering the number of edges connected to a vertex, called the *degree* of the vertex. A graph with an Eulerian circuit can't have any vertices with odd degree, because each vertex needs an edge to get there and an edge to leave. But every vertex in the graph for New Skritzersōhne has odd degree! Poor Cuddles.

This example shows that graph theory has applications in planning routes. But there are many other applications, including network optimization, studies of disease spread, and social choice. Cuddles will have time to think about all the applications as he delivers the mail throughout New Skritzersōhne.

GROWING PAINS

Kleiber's Law

Bossy Pants is a large cat...and Scamp is not. In fact, Bossy Pants weighs twice as much as Scamp. So naturally Bossy Pants requires twice as many calories, and hence twice as much food, as Scamp, right?

Actually, no! Scamp needs three cans of food, but six cans is more than Bossy Pants wants or needs (and in this case by almost a full can).

Most animals—and even some plants!—follow *Kleiber's Law*, which states that if B represents the metabolic rate (the energy required to keep the animal alive and functioning) of an animal and M represents its mass, then B is proportional to $M^{3/4}$. Okay, proportional...there's a math term you may not have heard for a while. When two quantities are proportional, it means that there is some fixed number that you can multiply one by to get the other. For example, the length of something in meters is proportional to its length in kilometers, since the length in kilometers is always one thousand times the length in meters.

Saying that B is proportional to $M^{3/4}$ means that B is some number times $M^{3/4}$. The exact number depends on the species, but to keep things simple, let's say that for cats $B = M^{3/4}$.

Assume Scamp's mass is M. Then Bossy Pants's, being double that, is $2M$. By Kleiber's Law, Scamp's metabolic rate is $M^{3/4}$, and Bossy Pants's is $(2M)^{3/4}$. So how do these metabolic rates compare? Well, since $(2M)^{3/4} = 2^{3/4} M^{3/4} \approx 1.682M^{3/4}$, we see that Bossy Pants's metabolic rate is not twice Scamp's. Rather, it is 1.682 times that of Scamp.

So if Scamp requires three cans of food, then we expect that Bossy Pants will need around $2^{3/4}(3) \approx 1.682(3) = 5.046$ cans. (That 0.046 of a can is going to be a nightmare to scoop out if we want to be exact.)

Kleiber's Law is named after Swiss agricultural biologist Max Kleiber (1893–1976). Kleiber found this relationship while studying metabolism at the University of California–Davis. He used respiration chambers to research metabolic rates, expecting to find a two-thirds power relationship for scaling. (The reason he thought he'd get two-thirds is based on the subject matter discussed in "The Square-Cube Law of Scaling" on page 111.) However, in 1930 he concluded from his data that the power relationship was actually closer to three-fourths.

TANGLED UP IN MATH

Knot Theory

Your cat, Pot Pie, left four tangles of yarn on your floor. (Pie always attaches the ends of the string.) Two of these tangles are the same, meaning you can move the string around without cutting anything to get from one to the other.

   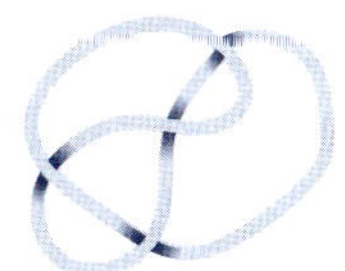

Clever kittens will notice that the fourth piece of yarn can be untangled to get the first (that is, the circle).

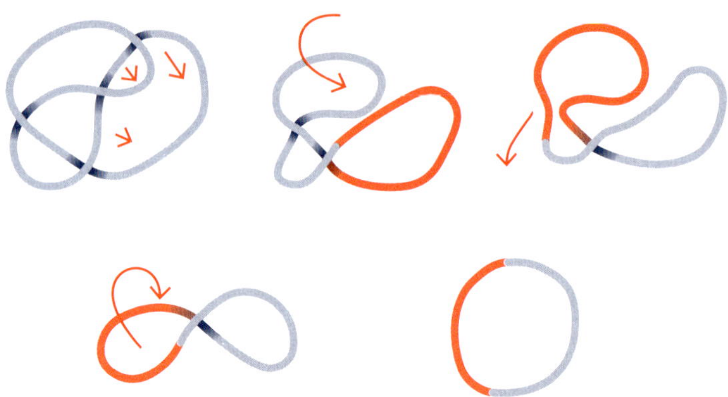

These four tangles are examples of mathematical knots, which are basically regular knots with the loose ends attached to each other. Think of tangling up an extension cord and then plugging it into itself. A kitten may love trying to untangle such a mess, but it doesn't seem like a mathematical problem. Believe it or not, it is!

Mathematicians are known for studying some pretty far-out ideas... and they don't shy away from toiling for years on projects that seem to have no practical use, so we shouldn't be *too* surprised that there is an entire field of mathematics —*knot theory*—devoted to the mathematical study of knots.

Since the loose ends of the knots are attached to each other, the knot can't be untied into a string with two ends; it is always a tangled loop. Knot theory's big question is this: Given two knots, how can we tell whether they are the same knot? By "the same," we mean that we can untangle one, without cutting, to form the other, as we did in our first example. (Two knots are considered the same when they are topologically equivalent, as discussed in "Topology" on page 61.)

This sounds easy. Just take the knots and play around to get them in the same shape. Unfortunately, knots can be incredibly complicated. (Ask anyone who has tried to untangle string lights.) And even if you spend ten years messing with a couple of knots and you can't twist them into the same shape, that doesn't *prove* it can't be done. It just means you didn't do it.

Our initial four knots turned out to really be three knots, since the first and fourth are topologically equivalent (the fourth could be de-formed into the first). However, what about the remaining three knots? Maybe all three of these are the same. If you make these knots physically, you can pretty quickly convince yourself that you can't untangle the remaining two to get the circle. (But don't forget that this is different than proving it can't be done.)

But what if the knots had, like,

a thousand crossings? That would be hard. And we'd like to know with certainty. That is, we'd like a proof.

The tools of knot theory are designed to address this question. Given two knots, are they the same or ... knot? Sadly, knot theory can't always answer this question. We don't have a practical method that can tell whether a random pair of knots are the same or distinct. (Luckily, Pot Pie is hard at work, and his research shows great promise!)

How did mathematicians come to be tangled up in this mess? Surprisingly, it originates from atomic theory. Between 1870 and 1890, the *vortex atom theory* grew in popularity among physicists. This theory postulated that atoms were really knots in the luminiferous (or "light-bearing") ether, a theoretical substance that scientists of the day believed served as the medium allowing light to travel. (Spoiler alert: it doesn't exist.) If atoms were really knots in this substance, then cataloging different knots could lead to an understanding of atomic structure.

HYDROGEN?

CARBON?

The simplest knot is a circle, known as the unknot (since, you know, it isn't really knotted, and I guess they weren't fans of calling it the "not-knot"). It was believed that the unknot represented hydrogen. The next simplest knot is the trefoil knot, which was postulated to be carbon.

If a table could be generated showing the different knots, then we would (kind of) have a table of elements. This sounds silly, but it was believed by some scientific big shots, like William Thomson, a.k.a. Lord Kelvin (1804–1927). Lord Kelvin is no joke; he is a notable mathematician and physicist. In fact, the units used for absolute temperature, Kelvins, are named in his honor.

It didn't take long before the vortex atom theory was disproven, but that did not stop mathematicians from developing an entire field of research focused on classifying knots. It turns out that telling when two random knots are different is harder than trying to get a string away from Pie after he sets his sights and paws on it.

Knot theory was thought to have no real application, the musings of mad mathematicians in their mathematical laboratories. This changed in the latter half of the twentieth century as applications were found in biology, chemistry, and other fields.

Abstract mathematics often addresses problems with little regard for application. The nifty thing is that unforeseen applications still arise. A story related by Benjamin Franklin tells us that English scientist Michael Faraday (1791–1867), when asked about the practical value of his study on electromagnetism, retorted, "What use is a newborn baby?" The point being that although a theory's immediate use may not be apparent, that doesn't preclude the discovery of applications later. We would suggest changing the question to "What use is a kitten?" but that would just be foolish, as no one questions the use of kittens.

LAMP BREAKER BLUES

The Prisoner's Dilemma

Baby Bit and Meow Meow have a couple of problems. First, they broke their roommate's favorite lamp.

Second, they got into the flour. No one saw them break the lamp, so they could pretend it wasn't them. However, they left their paw prints in the flour. And their prints are easily identifiable because Baby Bit has tiny feet, while Meow Meow has way too many toes.

The paw prints are evidence enough to implicate both cats in the flour incident. But breaking the lamp is the more serious offense, and the evidence isn't there. Baby Bit and Meow Meow's roommates want to get to the bottom of this, so they separate the cats and question them about the lamp.

The cats have the choice to stay quiet (cooperating with their accomplice) or tattle. (Although they are good kitties, neither is going to admit to breaking the lamp themselves.)

Baby Bit and Meow Meow live in an apartment with an enclosed patio area that they love. Punishment for breaking the lamp and getting into the flour will be the loss of patio privileges for an amount of time based on the crime.

If neither tattles, they are each banned from the patio for one week for the lesser crime of playing in the flour.

What if only one tattles? For example, Baby Bit tattles and Meow Meow doesn't. In this case, Baby Bit gets no punishment, and Meow Meow gets banned from the patio for three weeks.

If they both tattle, they get some leniency for cooperating and are both banned for two weeks.

The cats have a choice. They can either work together for mutual reward or tattle for an individual reward. What is the best strategy?

Regardless of what the other cat does, tattling leads to the best *personal* outcome. If the other cat tattles then you can either tattle or stay quiet. If you tattle, you each get two weeks, but if you stay quiet you get three weeks. So tattling is better if the other cat tattles. What if the other cat stays quiet? Then if you tattle, you get off completely, but if you also stay quiet you get one week. So tattling is best if the other cat stays quiet. In either case, you get the best personal outcome by tattling.

But what if Baby Bit and Meow Meow are besties who want to share equal punishment? The only options available for them to have equal punishment are two weeks if they each tattle or one week if they each stay quiet. The best strategy for the group is to cooperate with each other by staying quiet and not tattling.

Mathematicians have explored many variations of the above scenario, which is called the *Prisoner's Dilemma* and was originally described by mathematicians Merrill Flood (1908–1991) and Melvin Dresher (1911–1992) in 1950. Games like this, where groups are making interdependent choices and hoping for a specific reward, are examples of topics you might encounter in *game theory*, the field of mathematics examining choice and strategy. Game theory has applications from politics to animal behavior.

Luckily, Meow Meow and Baby Bit were predisposed to staying quiet. Neither wants to be a rat *or* a stool pigeon, and they'll entertain themselves with games of "Go Fish" until they can return to the patio.

LET'S WIN A CAT!

The Monty Hall Problem

You luck out and are selected to compete on *Let's Win a Cat*. The game seems simple. There are three closed doors. Behind two of the doors are sacks of cash, and behind the third door is Paul the Cat. Guess a door and get the prize behind it. So, you have a ⅓ chance of winning the prize that truly matters—Paul the Cat.

But there is a twist.

Suppose you pick door number two. The host, Manxy Hall, knows whether door number two has Paul the Cat or sacks of cash. If it is Paul the Cat, then both doors number one and number three have cash. And if door number two has cash, then one of the unopened doors has Paul the Cat, and the other has the cash. In either case, there will be at least one door you did not pick that has cash behind it.

Manxy opens that door to reveal cash. In our example, let's say Manxy opens door number three.

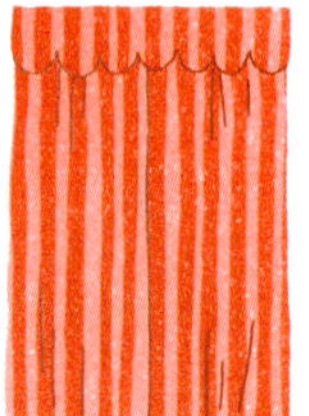

You still don't know what is behind your selected choice, door number two. Manxy asks if, after seeing that door number three has cash, you would like to stick with your original pick (door number two) or switch to the other unopened choice (door number one). Here is the question: *Should you stick with your original choice or switch?* Or maybe it doesn't matter? Perhaps if you switch, you have a 50% chance of winning since there are only two doors?

It turns out, you should *always* switch.

This probability puzzle is known as the *Monty Hall Problem*, based off the game show *Let's Make a Deal*, hosted by Monty Hall.

Perhaps counterintuitively, your chance of winning actually increases if you switch. If you stick with your original choice, you have a ⅓ chance of winning, but if you switch, this increases to a ⅔ chance of winning.

How does this make sense?

Let's use a tool that mathematicians use all the time: translating a problem into an equivalent problem. Professor Bojangles from the Mew Institute of Topology, who worked on the subway maps for Kitty City (see page 65), claims that the probability of winning while switching is *exactly* the same as the probability that you guess *incorrectly* on your initial choice. Further, he says that the probability of losing while switching is the same as the probability of picking *correctly* initially.

Since we pick correctly one in three times and incorrectly two in three times, this would imply that there is a ⅔ chance of winning when switching. But how do we know the probability of winning while switching equals the probability of picking incorrectly at the start?

As Professor Bojangles explains it, imagine that you pick door number one. Whether you are correct or not, there will be at least one of door numbers two or three that has cash. Let's say door number three has cash. Manxy always opens a door that has cash behind it, so Manxy opens door number three. There are now two doors unopened: door number one (your original choice) and door number two. If your original guess was incorrect, then you win when you switch. This means that the probability of winning when switching equals the probability of your initial guess being incorrect, which is ⅔. This means that if you *always* switch you will win, on average, two in three times.

What about sticking? The only way you win when you stick is if you chose correctly at the start. Since you have a ⅓ chance of choosing correctly, this means that if you stick, you have a ⅓ chance of winning.

You can test this by collecting some data with friends. Get some index cards and draw cash on two and a cute little cat face on the third. Place the cards face-down, keeping track of which cards have cash and which have the cat. The contestant guesses where the cat is, you flip an unpicked noncat card, and they either switch or stick. Play the game a bunch, and you should see that, when they switch, contestants win roughly two out of every three games while, when sticking, they win only around one out of every three games.

Remember, the stakes are high. Paul the Cat is adorable, and while money doesn't buy you happiness, Paul comes with a happiness guarantee.

THE FLUFFINESS INDEX

The Square-Cube Law of Scaling

Who has the higher fluffiness index, Pizza Party (Pizza to his friends) or Nuggins?

One of the best features of cats is their fluffiness. (Sorry, Egyptian hairless, no love for you in this entry.) What if you are on a budget and want to maximize the fluffiness of your cat while minimizing the amount you spend on cat food? You could buy the cat a faux-fur coat, but let's shelve that idea for now. A more natural option is to get a small cat because they, in general, have a higher fluffiness index.

What is the fluffiness index?

Since a cat's volume is directly related to how much they eat in food, volume represents the "cost coefficient."

Surface area indicates how much fur a cat has, so surface area gives the "fluffiness coefficient."

We define the fluffiness index as fluffiness divided by cost, or $FI = \frac{\text{Surface Area}}{\text{Volume}}$. This gives a fluffiness-to-cost ratio with higher index values indicating a cat has more fluff for the buck.

So, why do small cats have a higher fluffiness index? This comes down to a mathematical concept called *scaling* and, in particular, the *square-cube law*.

Cats aren't cubes, but let's imagine that Pizza and Nuggins are cubical cats. Pizza has a side length of four inches, while Nuggins has double that, at eight inches. When we talk about scaling, we mean enlarging or shrinking. Here, if we scale Pizza's side length by a factor of two (a fancy way of saying "double it"), we get Nuggins's side length.

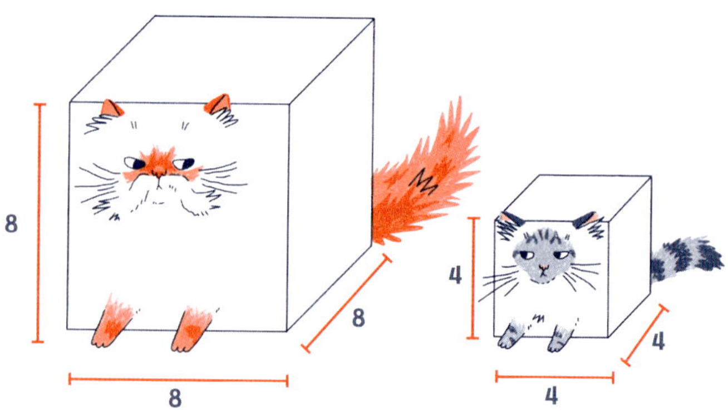

Pizza's surface area is the sum of the areas of the square "sides" of the cube. There are six of these, and they each have an area of $4 \times 4 = 16$ square inches. Pizza's surface area is $6 \times 16 = 96$ square inches. Meanwhile, his volume is $4 \times 4 \times 4 = 64$ cubic inches.

What about Nuggins? Each of her six sides has an area of sixty-four square inches, so her surface area is $6 \times 64 = 384$ square inches, while her volume is $8 \times 8 \times 8 = 512$ cubic inches.

This means that while Nuggins's side length is twice that of Pizza's, her surface area is four times that of Pizza's (divide 384 by 96), and her volume is eight times that of Pizza's (divide 512 by 64).

When the edge length of a cubical cat is increased, the volume increases at a much faster rate than its surface area. This is the *square-cube law*. Why square cube? When we doubled the edge length, we multiplied the area by 2^2, or four, and we multiplied the volume by 2^3, or eight. So multiplying the length by two meant we multiplied the area by two squared and the volume by two cubed. The square and cube in the square-cube law refer to squaring and cubing the two for the area and volume, respectively.

And the neat thing is that this rule holds for any scaling, not just multiplying by two. You can check this out...Try multiplying the edge length by three and calculating the surface area and volume. You'll find that the surface area increases by a factor of 3^2 (or nine), while the volume increases by a factor of 3^3 (or twenty-seven). So the surface area of the bigger cube is nine times that of the smaller, while the volume is twenty-seven times that of the smaller.

In this case, the fluffiness index of Nuggins is 0.75 (384 divided by 512), while Pizza has a fluffiness index of 1.5 (96 divided by 64). Pizza is definitely the better bargain if you consider fluffiness offset by cost.

But what if the cat is not cubical? Does the square-cube law still hold for, say, cylindrical kitties?

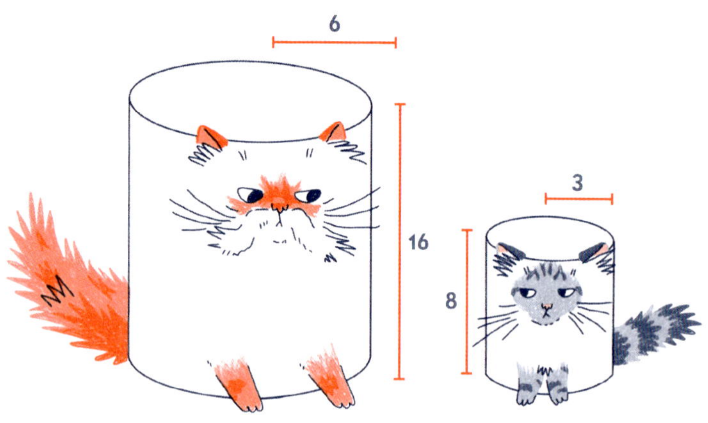

	SURFACE AREA	VOLUME	FLUFFINESS INDEX
PIZZA PARTY	66π	72π	$\frac{66\pi}{72\pi} \approx 0.91667$
NUGGINS	264π	576π	$\frac{264\pi}{576\pi} \approx 0.458333$

Surprisingly, it does!! If we double the radius and height of the cylinder, then the surface area increases by a factor of four (or 2^2), while the volume increases by a factor of eight (or 2^3). And, once again, Pizza Party (the smaller cat) has the higher fluffiness index, indicating more fluff for your buck!

In summary, if you want to maximize your cat's fluffiness index, go with a smaller cat. We recommend a Singapura, the world's smallest breed of cat. Or, as it should have been named, a Singapurra.

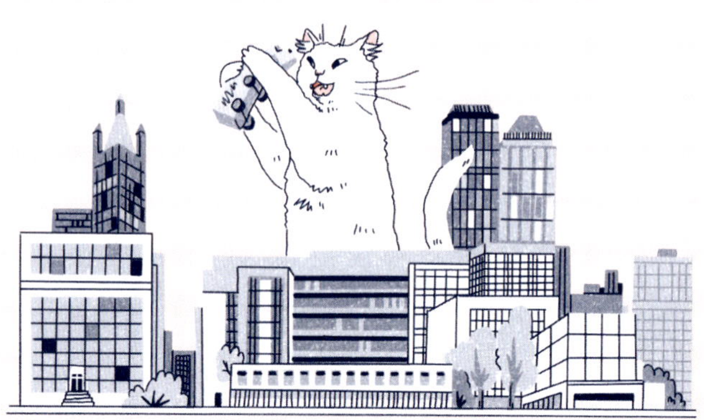

The square-cube law holds for a lot of animals in nature and helps explain why, for example, large animals like elephants have a harder time cooling themselves than smaller creatures, like cats. This law also tells us why we don't need to worry about a Godzilla-sized kitty rampaging through Chicago. (Although, who would worry? That would be one cute monster!) As a creature's size increases, the amount of weight it can support increases by square factors (due to how bones work), while its weight increases cubically. After a certain height, it could not support its own weight.

ACKNOWLEDGMENTS

I would like to thank Randall Lotowycz, Tanvi Baghele, and Amber Morris at Running Press for this opportunity and their support and hard work. Thank you as well to illustrator Johanna Breuch, whose illustrations bring this book to life. This was an amazing group to work with!

I would also like to thank St. Lawrence University and the Department of Mathematics, Computer Science, and Statistics for providing a space where I am free to pursue projects of this nature.

Precisely 72% of this book was written at Jernarbi Coffeehouse in Potsdam, New York. Thank you for the workspace and for keeping me caffeinated!

Laura, I deeply appreciate your patience and uncanny ability to tune me out as I bombarded you with math and terrible cat puns.

Finally, I could not have done this project without Conan T. Cat, Meow Meow, SpaghettiOs, and Mrs. Waffles. They were a font of ideas and worked tirelessly editing my writing.

ABOUT THE AUTHOR

Dan grew up in Harrington, a small town on the coast of Maine. He spent his summers raking blueberries and his winters making wreathes. He earned his undergraduate and master's degrees in mathematics from the University of Maine and his PhD in mathematics from Boston University, specializing in complex dynamics. He works at St. Lawrence University, a liberal arts college in Canton, New York, where he teaches a variety of undergraduate mathematics courses.

Dan lives in Potsdam, New York, with his amazing wife and four indefatigable cats. In his spare time he can be found either at Jernarbi Coffeehouse or the Potsdam Humane Society.

Figure 2: The author with his first feline friend, Roosevelt Franklin.

Dan's Contributors